PLANT TISSUE CULTURE

PLANT TISSUE CULTURE
Practices and New Experimental Protocols

B.N. SATHYANARAYANA
Professor
Division of Horticulture
University of Agricultural Sciences
Bangalore

DALIA B. VARGHESE
Post Doctoral Fellow
School of Biological and Conservations Sciences
University of Kwa Zulu Natal
South Africa

TECHSAR™
NEW DELHI

Published and Printed by
Techsar Pvt. Ltd.
4435–36/7, Ansari Road, Daryaganj
New Delhi-110002 (India)
E-mail: tech_info@techsarworld.com
Website: www.techsarworld.com

ISBN 978-81-89866-11-2

Reprint 2025

Dedication

The book is dedicated to my beloved wife Shashi,
and loving children Appu and Bhavvi,
for their love and encouragement.

B.N. Sathyanarayana

Foreword

The major steps in the development of plant tissue culture have been made 40 to 70 years ago. Growth under sterile conditions on artificial nutrient medium was achieved in the 1930s. The requirements for inorganic and organic nutrition, and for plant growth regulators were established in the 1950s and 1960s. During the same period, adventitious shoot formation, somatic embryogenesis, axillary branching, protoplast culture and haploid production were achieved. Plant tissue culture has become an essential tool in breeding and propagation of many crops and also in basic scientific research. Surprisingly, there are only few books presenting a comprehensive introduction into the practice of plant tissue culture. This book of Dr. Sathyanarayana makes a good effort to fulfill the deficiency in an excellent way. It presents an introduction from the point of view of a scientist and teacher who spends every day in the laboratory. In this book, Dr. Sathyanarayana combines his extensive knowledge on tissue culture with a broad experience about the needs of students. The chapters deal with the various biological technologies in plant tissue culture like meristem culture, axillary branching, protoplast culture and haploid production, and with the technological aspects such as medium preparation, sterilization, equipment and laboratory design. The chapters deal with laboratory essentials and clearly describe how to carry out the protocols. The book is written by an author who knows what has to be stressed. Dr. Sathyanarayana emphasizes, for example, the need for a checklist when preparing media. Every researcher who frequently prepares tissue culture media has surely doubted occasionally whether or not he has forgotten something and poured the medium through the sink to be absolutely safe. Ticking the successive actions on a checklist immediately after completion is the best solution for this problem as

Dr. Sathyanarayana explains in a narrative way. But the book also shows results of research such as illustrative fluorescence photographs of developing root initials. The organization in well-chosen topics for each chapter makes the book an excellent book of reference. This book is an essential part of the bookshelf of all researchers that use plant tissue culture.

Dr. Geert-Jan de Klerk,
Wageningen Tissue Culture Center,
Wageningen
The Netherlands
Editor-in-Chief
Plant Cell, Tissue and Organ Culture

PREFACE

Technologies have been undergoing a sea change over centuries. Arguments prevail whether these changes have been faster now, or during earlier centuries. In our opinion, these arguments remain acceptable with people's bent of mind over time to time.

Biotechnology is a buzzword in the field of science. It has its roots in the inventions and innovations, which took place over millennia and manipulated organisms to regulate and improve everything that comes under the realm of necessity for the human welfare and the welfare of plethora of organisms, whether higher or lower, serving the human kind. But, what is relevant now, concerning those who are living at present and those who are going to be a part of this globe in future, are the trends taking place in the technologies surrounding us, and which are being acid tested, used, and improved. Hence, apart from other technologies, biotechnology is by far the most relevant technology. It touches the core of our existence and the need for our survival; controversies not withstanding.

Plant tissue culture forms the basis of present day biotechnology. The techniques of tissue culture are exciting academically as well as for commercial production. There is a dire need to have a book on plant tissue culture, especially *focused on the aspects of practical procedures and protocols of tissue culture.* The students always feel the unavailability of books on such subjects, which are simpler in their core and contents. Hence, this attempt is to fill the void of such necessity. *This book contains important experimental protocols that provide guidance on carrying out hands-on experiences. It gives all the procedures as a general guideline, and also necessary tips to maneuver any problem encountered. These tips are based on our own experiences while teaching and researching the techniques of plant tissue culture.*

This book is unique and stands out when compared with other books of similar contents, for the reason that *many techniques are demonstrated describing the actual protocols experimented and developed with many plant species by different scientists.* We are sure that the readers of this book, the students, the researchers, and industrialists would find it interesting.

B.N. Sathyanarayana

ACKNOWLEDGEMENTS

Though, dreamt over a decade by me to write a book on plant tissue culture, it has just materialized now. It is good for the contents of the book that this project took so many years to take off. The book hence, is now richer with my experiences meld into it.

The first author would like to remember his close friends Mr. D.P. Prakash, the former Director of Jain Institute of Vocational and Advanced Studies (JIVAS), Bangalore, whose encouraging teasers has made him complete the effort; Dr. S. Mahesh of Bangalore University for his words of encouragement and inspiration to embark upon this book; postgraduate students of plant tissue culture in the Faculty of Horticulture; in this Ms. Meera Manjusha, a Ph.D. student, should find a special mention in the preparation of the manuscript by her keen coordination with the second author. Later, Mr. Hanumantharaya, a doctorate student, assisted in every manner in correcting manuscript, reading the manuscript and also compiling bibliography and contents.

Words of appreciation and encouragement to realize the potential of the first author, by Dr. P.G. Chengappa, the former Director of Instruction of Agricultural College, GKVK (presently the Vice-chancellor of the University of Agricultural Sciences, Bangalore) is sweetly remembered here. I also wish to thank Dr. A.A. Farooqi, Former Professor and Head, for his constant persuasion to write a book on plant tissue culture.

I owe a lot to Dr. Jennet Blake, my Supervisor of doctoral studies at the University of London, Wye College (now Imperial College), for nuances of plant tissue culture introduced to me by her.

I thank Dr. Geert-Jan de Klerk, the Editor-in-Chief, Plant Cell Tissue and Organ Culture, Senior researcher, Plant Tissue Culture, Wageningen Tissue Culture Centre, Biodiversity and Breeding, Plant Research International, Netherlands, for having been generous enough to write a foreword to this book.

Words of encouragement, friendship and good wishes of my collegues of the Department of Horticulture, is greatly acknowledged.

I am greatly indebted to my University (University of Agricultural Sciences, Bangalore) for giving me everything to realize my potential.

B.N. Sathyanarayana

CONTENTS

1. Historical Developments in Plant Tissue and Cell Culture **1**
2. Laboratory Design and Development **9**
2.1. The Building 9
2.1.1. Structures and Utilities 9
2.1.2. Layout 10
2.2. The Rooms 10
2.2.1. Media Preparation Area 11
2.2.2. Washing/Sterilizing Area 12
2.2.3. Inoculation Room/Sterile Transfer Room 12
2.2.4. Culture Room/Incubation Room/Growth Room 13
2.2.5. General Storage Area 16
2.2.6. Research Area 16
2.2.7. Planting and/or Shipping Area 17
3. Instrumentation **19**
3.1. Laminar Flow Hoods 19
3.1.1. Operation and Management 20
3.1.2. Caution while using Laminar Flow Cabinets, Particularly Bunsen Burners and UV Lights 21
3.2. Autoclaves 22
3.2.1. Operation and Management 23
3.2.2. Caution 23
3.3. Distillation Unit 24
3.3.1. Reverse Osmosis Water 24
3.3.2. The Millipore System 24
3.3.3. Operation and Management 25
3.4. Digital pH Meter 25
3.4.1. Operation and Management 25
3.5. Orbital Shaker 26
3.5.1. Operation and Management 26

3.6. Inverted Microscope 26
3.7. Other Instruments 27
3.8. Culture Vessels 27
4. Plant Tissue Culture Media 29
4.1. Solid and Liquid Media 30
4.2. Media Composition 32
4.2.1. Inorganic Nutrients 32
4.2.1.1. Macronutrients 32
4.2.1.2. Micronutrients 33
4.2.2. Organic Supplements 35
4.2.3. Growth Regulators 35
4.2.4. Source of Carbon and Energy 37
4.2.5. Solidifying Agents 38
4.2.6. Miscellaneous Constituents 40
4.2.7. Types of Media 44
4.2.8. Preparation of Plant Tissue Culture Media 46
4.2.9. Common Problems Encountered 51
4.2.10. Preparation of 1 L of MS Medium (full strength) from Stock Solutions 51
4.2.11. Preparation of Half Strength MS Media (1 L) from Stock Solutions 52
5. Sterilization Techniques 55
5.1. Equipments Used for Sterilization 55
5.1.1. Sterilization of Culture Vessels and Instruments 56
5.1.2. Sterilization of Nutrient Media 57
5.1.3. Sterilization of Explants 58
5.1.4. Sterilization of Transfer Area 60
5.1.5. Sterile Transfer 60
5.2. Contamination of Plant Tissue Cultures 62
5.3. Control of Contamination by the Use of Antibiotics 62
5.4. Recovery of Contaminated Cultures 63
6. Safety Rules and Regulations 65
7. Applications of Plant Tissue Culture 67
7.1. Micropropagation 68
7.2. Crop Improvement 70
7.2.1. Anther and Pollen Culture 70
7.2.2. Callus and Suspension Cultures 70
7.2.3. Protoplast Culture 71
7.2.4. Transformation 71
7.2.4.1. Some Methods for Transferring Foreign DNA into Plant Cells 71

7.2.4.1.1. *Agrobacterium*-mediated Transformation 72

7.3. *In Vitro* Germplasm Conservation 73

7.4. Secondary Metabolite Production 73

8. Micropropagation 79

8.1. Micropropagation by the Release and Growth of Pre-existing Meristems 80

8.2. Micropropagation by Adventitious Shoot Proliferation 82

8.3. Plant Regeneration by Direct Somatic Embryogenesis 83

8.4. Plant Regeneration by Indirect Somatic Embryogenesis 83

8.5. Applications of Micropropagation 84

8.6. Stages of Micropropagation 84

9. Micropropagation by Proliferation of Axillary Buds 87

9.1. Shoot Tip Culture 87

9.2. Single Node Culture 87

9.3. Common Problems Encountered 90

9.4. Protocol for Micropropagation of Brahmi by Axillary Bud Proliferation 92

9.5. Protocol for Micropropagation of Curry Leaf from Nodal Segments 93

9.6. Protocol for Micropropagation of Jackfruit 95

9.7. Protocol for *in vitro* Propagation of Coleus 96

9.8. Protocol for Direct Shoot Regeneration from Shoot Tips of Gerbera 98

10. Micropropagation by Adventitious Shoot Regeneration 101

10.1. Objectives 101

10.2. Equipments 101

10.3. Media and Reagents 101

10.4. Plant Materials 102

10.5. Procedure 102

10.5.1. Culture Initiation 102

10.5.2. Shoot Multiplication 102

10.5.3. Rooting of Shoots 102

10.5.4. Plantlet Establishment 102

10.6. Schedule of Observations 103

10.7. Common Problems Encountered 103

10.7.1. Brown/Black Explants 103

10.7.2. Lack of Shoot Multiplication 103

10.8. Protocol for Micropropagation of Brahmi by Adventitious Shoot Formation 104

10.9. Protocol for Micropropagation of Patchouli by Adventitious Shoot Formation 106
10.10. Protocol for Direct Adventitious Shoot Formation in Carnation 106
11. Plant Regeneration from Callus Cultures 109
11.1. Measurement of Callus Growth 110
11.2. Objectives 110
11.3. Equipments 110
11.4. Media and Reagents 110
11.5. Plant Materials 111
11.6. Procedure 111
11.6.1. Aseptic Germination of Seeds 111
11.6.2. Callus Initiation and Maintenance 111
11.6.3. Induction of Shoots from Callus 111
11.7. Schedule of Observations 112
11.8. Common Problems Encountered 113
11.9. Protocol for Micropropagation of Brahmi by Indirect Organogenesis 113
11.9.1. Surface Sterilization and Explant Preparation 113
11.9.2. Culture Initiation and Maintenance 113
11.9.3. Shoot Regeneration from Calli 113
11.9.4. Rooting and Establishment of the Plantlets 113
11.10. Protocol for Callus Induction and Shoot Regeneration in Carnation 115
11.11. Protocol for Callus Induction in Chilli 116
11.12. Protocol for Callus Induction and Shoot Regeneration in *Anthurium andreanum* Lind. 117
11.13. Protocol for *in vitro* Propagation of Coleus by Indirect Organogenesis 119
11.14. Protocol for Callus Induction and Shoot Regeneration in Gerbera 120
12. Plant Regeneration by Somatic Embryogenesis 123
12.1. Objectives 124
12.2. Equipments 124
12.3. Media and Reagents 124
12.4. Plant Material 125
12.5. Procedure 125
12.6. Common Problems Encountered 125
12.6.1. Abnormal Somatic Embryos 125
12.7. Synthetic Seeds/Artificial Seeds 126
12.7.1. Advantages of Artificial Seeds 126

12.7.2. Protocol for Encapsulation of Seeds 126
12.7.2.1. Gel Complexation via Dropping Procedure 126
12.7.2.2. Alternate Method 126
12.8. Protocol for Somatic Embryogenesis and Plantlet Regeneration from Node Callus Cultures of *Gloriosa superba* L. 127
12.9. Protocol for Direct Somatic Embryogenesis in Sweet Type Carambola 129
13. Meristem Culture **131**
13.1. Objectives 132
13.2. Equipments 132
13.3. Media and Reagents 132
13.4. Procedure 132
13.5. Problems and their Solutions 133
13.5.1. Poor Growth of Meristems 133
14. *In Vitro* Haploid Production **135**
14.1. Uses and Applications of Anther Culture 135
14.2. *In Vitro* Methods for the Induction of Haploids 137
14.3. Androgenic Haploid Production 137
14.4. Anther Culture 137
14.5. Ontogeny of Androgenesis 138
14.6. Direct Androgenesis 139
14.7. Indirect Androgenesis 139
14.8. Culture Media and Nutritional Requirements 139
14.9. Culture of Isolated Pollen Grains/Pollen Culture 140
14.10. Methods of Pollen Culture 141
14.10.1. Hanging Drop Method of Pollen Culture 141
14.10.2. Nurse Culture Technique 141
14.10.3. Pre Culture Method 141
14.11. Precautions to be Considered during Culturing of Pollen 142
14.12. Culture of Haploid Protoplast 142
14.13. Factors Affecting Androgenesis 143
14.14. Stability of Haploids 144
14.15. Gynogenic Haploid Production 145
14.16. Limitations of *in vitro* Haploid Production 145
14.17. Anther Culture 145
14.18. Ovary and Ovule Culture 146
14.19. Problems Associated with *in vitro* Haploid Production 147
15. Embryo Culture **149**
15.1. Embryo Culture and its Applications 149

15.2 Excision of Embryos 150
15.3. Types of Embryo Culture 150
15.3.1. Culture of Immature Embryos 150
15.3.2. Culture of Mature Embryos 150
15.4. Common Problems Encountered 152
16. Endosperm Culture **153**
16.1. Applications of Endosperm Culture in Horticulture 154
16.2. Objective 154
16.3. Equipments 154
16.4. Media and Reagents 155
16.5. Procedure 155
16.6. Problems Encountered 156
17. Culturing of Roots *In Vitro* **157**
17.1. Objective 158
17.2. Materials Required 158
17.3. Media and Reagents 158
17.4. Procedure 158
18. Protoplast Isolation and Culture **161**
18.1. Protoplast Isolation 161
18.1.1. Sources of Material 161
18.1.2. Methods of Isolation 162
18.1.2.1. Use of Osmoticum 162
18.1.2.2. Mechanical Isolation of Protoplasts 162
18.1.2.3. Enzymatic Isolation 163
18.1.3. Protocol for Protoplast Isolation from Leaves 163
18.1.3.1. Preparation of the Explant 163
18.1.3.2. Protoplast Isolation by Enzymatic Method 165
18.1.3.3. Purification of the Protoplasts 165
18.2. Testing the Viability of the Protoplasts 166
18.3. Culture of Protoplasts 166
18.3.1. Agar Embedded Culture 167
18.3.2. Microchambers 167
18.3.3. Hanging Drop Culture Technique 167
18.4. Regeneration of Cell Wall by the Cultured Protoplasts 167
18.5. Regeneration of Plantlets 168
18.6. Protocol for Protoplast Isolation from Carrot 168
18.7. Isolation of Protoplasts from Leaves of Chilli 169
18.8. Isolation of Protoplasts from Leaves of *Anthurium andreanum* Lind. 170

19. Protoplast Fusion and Cybridization **173**
19.1. Cybridization 173
19.2. Protoplast Fusion 174
19.2.1. Spontaneous Fusion of Protoplasts 174
19.2.2. Induced Fusion of Protoplasts 174
19.2.2.1. Chemical Fusion 174
19.2.2.2. Electrofusion 174
19.3. Selection of Protoplast Fusion Hybrids 175
19.3.1. Visual Selection 175
19.3.2. Fluorescent Labels 175
19.3.3. Drug Sensitivity and Resistance 175
19.4. Procedure for Chemical Induced Fusion of Protoplasts 175
20. Storage Organ Culture **179**
20.1. The Production of Bulbils and Cormlets 179
20.2. Protocorm Formation by Orchids 179
20.3. Miniature Tubers 180
20.4. *In Vitro* Tuberization in Potato 180
20.5. Advantages of using *in vitro* Microtubers 180
20.6. Protocol for *in vitro* Microtuber Production in Potato 180
21. *In Vitro* Germplasm Conservation **183**
21.1. Advantages of *in vitro* Germplasm Conservation 184
21.2. *In vitro* Germplasm Exchange 184
21.3. Theoretical Basis of *in vitro* Germplasm Conservation 184
21.4. Transfer of Water to the Solid State 184
21.5. Transfer of Water to Gaseous State 185
21.6. The Substitution of Water 185
21.7. Plant Material used for *in vitro* Conservation 185
21.8. *In vitro* Preservation of Seed Propagated Species 185
21.9. *In vitro* Preservation of Vegetatively Propagated Species 185
21.9.1. Maintenance of Cultures at Normal Growth Conditions 185
21.9.2. Maintenance of Cultures at Minimal Growth Conditions 186
21.9.3. Cryopreservation 186
21.9.3.1 *In vitro* Preservation of Pollen 186
21.9.3.2 Preservation of Somatic Embryos 187
21.10. Molecular Techniques of Germplasm Conservation 187
21.11. Stages of Cryopreservation 188
21.11.1. Pre-growth Stage/Pre-treatment Stage 188
21.11.2. Cryoprotection 188

21.11.3. Cooling and Storage 189
21.11.3.1. Two-step/Slow /Optimum Rate Cooling 189
21.11.3.2. Rapid Cooling/Vitrification 189
21.11.3.3. Thawing 189
21.11.3.4. Recovery of Cultures 189
22. *In Vitro* Production of Plant Secondary Metabolites 191
22.1. Requirements for Successful *in vitro* Secondary Metabolite Production 192
22.2. Plant Cell Cultures 192
22.2.1. Immobilized Cell Culture 193
22.3. Culture of Differentiated Organs 194
22.3.1. Hairy Root Culture 194
22.3.1.1. Characteristics of Hairy Root Cultures 194
22.3.1.2. Secondary Metabolite Production by Hairy Roots 194
22.3.1.3. Scaling-up of Hairy Roots 195
22.4. Biotransformation 195
22.5. Stimulation of Chemical Production 195
22.5.1. Elicitation 195
22.5.2. Precursor Feeding 196
23. Acclimatization of Micropropagated Plantlets 197
23.1. Water Stress 197
23.2. The Cuticle 198
23.3. Stomata 198
23.4. Hydathodes 199
23.5. Leaf, Stem and Root Anatomy of Micropropagated Plantlets 199
23.6. Photosynthesis 200
23.7. Techniques of Acclimatization 200
23.7.1. Procedures for Acclimatization 201
23.7.2. Humidity 201
23.7.3. Antitranspirants 201
23.7.4. Light 201
23.7.5. Soil and Containers 201
23.8. Control of Diseases 202
23.9. Control of Temperature 202
23.10. Limitations in Acclimatization Practices by Students 202
24. *In Vitro* and *Ex Vitro* Rooting of Microshoots 203
24.1. *In vitro* Rooting 203
24.1.1. Use of Growth Regulators for *in vitro* Rooting 204

24.1.2. Media Effects 204
24.1.3. Use of Activated Charcoal 204
24.1.4. Culture Environment 204
24.1.5. Protocol for *in vitro* Rooting of Jackfruit 205
24.2. *Ex vitro* Rooting 205
24.3. Anatomical Changes During Rooting 206
24.4. Protocol for *ex vitro* Rooting of *in vitro* Grown Microcuttings of Jackfruit 210
25. Somaclonal and Gametoclonal Variations 213
25.1. Sources of Somaclonal Variation 214
25.2. Factors Determining Somaclonal Variation 214
25.3. Genetic Mechanisms/Causes for Somaclonal Variation 214
25.4. Methods of Assessment of Somaclonal Variation 215
25.5. Applications in Plant Breeding 216
26. *Agrobacterium* Mediated Transformation 217
26.1. Delivery Systems 218
26.2. Biological Delivery System 218
26.3. The Ti Plasmid and Tumorgenesis 219
26.3.1. Host Recognition and Binding 219
26.3.2. Activation of the *Vir* Genes 220
26.3.3. T-Region Processing 220
26.3.4. Formation of the T-Complex 220
26.3.5. T-Complex Transfer to the Plant Cell 221
26.3.6. Integration into Host Genome 221
26.3.7. Tumor Formation 221
26.4. Limitations of Biological Delivery Systems 222
26.5. Current Developments and Techniques in Biological Delivery Systems 222
26.6. Integration, Recombination and Expression 224
26.7. Selectable Markers and Reporter Genes 224
26.8. Antibiotic Markers 224
26.9. Herbicide Markers 225
26.10. Reporter Genes 225
26.11. Recovery of Viable Transgenic Plants 226
26.11.1. Selection 226
26.11.2. Regeneration 226
26.11.3. Reproduction 227
26.12. The Future of Plant Transformation 227
27. Problems in Plant Tissue Culture 229
27.1. Culture Contamination 229

27.1.1. Effects of Contamination Microorganism on *in vitro* Cultures 230
27.1.2. Organisms Associated with Plant Surfaces 230
27.1.3. Screening Plants for Contaminating Microorganisms 231
27.1.4. Procedures 232
27.1.4.1. Indexing Media and Preparation 232
27.1.5. Culture Indexing Procedure 233
27.2. Vitrification 234
27.3. Anatomy and Morphology of Malformed Plants *in vitro* 234
27.3.1. Morphological Characters 234
27.3.2. Anatomical Characters 235
27.4. Factors Influencing Hyperhydricity 235
27.5. Prevention of Hyperhydricity 236
27.6. Tissue Blackening or Browning 237
27.6.1. Factors Affecting Tissue Blackening or Browning 238
27.6.2. Causes of Browning 238
27.6.3. Prevention of Browning 238
27.7. Minimizing the Damage Caused to the Explant 238
27.8. Removing the Phenolic Compounds Produced 239
27.8.1. Leaching or Dispersal 239
27.8.2. Frequent Transfers 239
27.8.3. Use of Liquid Media 239
27.8.4. Adsorption with Activated Charcoal 239
27.8.5. Adsorption with Polyvinylpyrrolidone 240
27.9. Decline in Vigour 241
27.9.1. Decline During Initiation 242
27.9.2. Decline at Later Stages 242
27.10. Cytokinin Habituation 242
27.11. Shoot Necrosis 242
27.12. Causes for Apical Necrosis 242
27.13. Prevention of Shoot Necrosis 243
28. Commercial Micropropagation **245**
28.1. Designing a Laboratory 246
28.2. Facilities 246
28.3. General Laboratory Design 247
28.4. Media Preparation and Sterilization Area 249
28.5. Growth Room 250
28.6. Aseptic Transfer Area 252

28.7. The Cost of Micropropagation 252
28.8. Methods to Reduce Cost of Micropropagation 253
28.8.1. Reduction of Labour Costs 253
28.8.2. Choice of Containers 253
28.8.3. The Cost of Materials 254
28.8.4. Preparation for Planting and Dispatch 254
28.8.5. Increasing Plant Multiplication Rates 255
28.8.6. The Importance of Timing 255
28.8.7. Growth Room Utilization 255
28.8.8. Photoautotrophic Growth of Cultures 256
28.8.9. Mechanization and Automation 257
28.8.9.1. Automation of Plant Tissue Culture Techniques 257
28.8.9.1.1 Use of Robots 257
28.8.9.1.2 Use of Bioreactors 257
28.8.9.2. Types of Bioreactors 259
28.8.9.2.1 Stirred Tank Bioreactors (STB) 259
28.8.9.2.2 Airlift Bioreactors (ALB) 259
28.8.10. Alternative Cultural Techniques 260
28.8.10.1. Shoot Hedges 260
28.8.10.2. Aeroponic Culture 260
28.8.10.3. Direct Shoot Formation 260
28.8.10.4. Encapsulated Shoot Buds 260
28.8.10.5. Macerated Tissues 260
29. Microscopic and Photographic Techniques for Plant Tissue Culture 261
29.1. Microscopy 261
29.2. The Concept of Magnification 261
29.3. Lenses and Geometrical Optics 261
29.4. Microscope Optical Components 261
29.5. Microscope Illumination 262
29.5.1. Image Brightness 262
29.6. Microscope Objectives 262
29.7. Infinity Optical Systems 264
29.8. Eyepieces (Oculars) 264
29.9. Substage Condensers 265
29.10. Specimen Stages 265
29.11. Reflected Light Microscopy 265
29.12. Fundamentals of Stereomicroscopy 265
29.13. Photographic Techniques 266

29.14. Type of Photography for Plant Cell and Tissue Culture Research 266
29.14.1. Macrophotography 266
29.14.2. Microphotography 266
29.15. Procedure for Jackfruit Plant Sample Preparation 269
29.15.1. Materials Required 269
29.15.2. Procedure 269
29.15.3. Light Sources 270
29.16. Types of Films 270
29.16.1. Colour Slide Film 270
29.16.2. Black and White Film 270
29.17. Composition 270
30. Histological Techniques 271
30.1. General Considerations 271
30.1.1. Equipment and Supplies 271
30.2. General Protocol for Paraffin Studies 272
30.2.1. Fixation 272
30.2.2. Dehydration and Infiltration 272
30.2.3. Casting Specimens into Blocks 272
30.2.4. Microtomy and Mounting Sections on Slide 273
30.2.5. Staining of Sections 273
Glossary 275
Appendix 283
Selected Bibliography 285
Index 289

1

Historical Developments in Plant Tissue and Cell Culture

The universities in India have been engaged in the basic and applied research ever since the beginning of this century. Applied research, however, got a special boost when several institutes were set up to carry out specifically application-oriented research. These institutions emphasized not only applied research but they also developed laboratories for basic research as well. Therefore, since India's independence, a concerted effort has been made in the Indian laboratories for scientific advancements in the country and this has led to the creation of a large resource of trained scientific personnel.

To give this research a further boost, the National Biotechnology Board was established in the early eighties, which became a full-fledged Government Department of Biotechnology. Biotechnology can be defined best as a science in which biological systems or processes are exploited for the benefit of humankind. It is an amalgamation of all sciences such as physics, chemistry, mathematics and biology as well as engineering technology.

The history of biotechnology is the same as the history of humankind's desire for survival by controlling nature and life. For thousands of years, man has been involved in such activities as brewing, food preservation and modification of food by fermentation (cheese, vinegar etc.), the manufacture of perfumes from aromatic flowers and invention of primitive medications. It is, however, the discovery of genetic engineering techniques via recombinant DNA technology, which is responsible for the current biotechnology boom. Using laboratory skills, which were unheard of a generation ago, scientists have isolated, put together and manipulated genes and have cloned to create life itself. The work of several brilliant scientists around 1950 led to the acceptance of a chemical compound—deoxyribonucleic acid (DNA) - as the magic substance of the gene which controls all characters of an organism. The unravelling of the DNA double helix was one of the great events in science, comparable to the splitting of the atom or the publication of Darwin's "Origin of Species". It also marked the coming of the age of a bold new science-Molecular Biology. At present, this has become one of the most active, exciting and productive arenas of science that has cornered much of the limelight focused on scientific developments and naturally, has attracted some of the best talents.

Science and molecular biology entered Indian laboratories as early as 1960, when basic training was being imparted on the understanding of DNA structure and mechanism of its action in test tubes, mostly using bacterial systems. Today, the scientists in laboratories around the world are increasingly getting closer to giving man, the control over his life and destiny. It is a prospect of great hope but is also fraught with extreme danger of misuse.

Plant genetic engineering is another important area in biotechnology, which has progressed to a great extent in India. Several species have been genetically transformed by inserting and expressing useful foreign genes. The protocols of regeneration of a complete plant from a single cell or a protoplast or a group of tissues is a prerequisite for genetic engineering of plants. Also, appropriate and foolproof procedures for cloning, sequencing and gene targeting in the plant are also a must. Scientists at several laboratories in India, such as Jawahar Lal Nehru University, New Delhi, Bose Institute, Calcutta, National Botanic Research Institute Lucknow, have cloned and sequenced many useful genes that are involved in providing resistance to diseases as well as stress conditions. Besides genes for enhanced storage life, genes involved in amino-acid biosynthesis have also been cloned in J.N.U. and inserted in legumes, rice and potato.

The Department of Biotechnology has developed seven centre of Plant Molecular Biology for generating improved varieties of plants through unconventional breeding techniques. These are at the J.N.U., New Delhi, South Campus, Delhi University, NBRI, Lucknow, MKU, Coimbatore and Bose Institute, Calcutta. All the centres have produced excellent results within a span of seven years. In plant tissue culture, India has always been at the forefront as compared to the rest of the world.

A novel technique of test-tube fertilization was developed in Delhi University to overcome incompatibility in plants, which is exhibited in wide crossing. This technique, developed in the 1960's, is being employed in many laboratories in all parts of the world. Another landmark research in plant breeding and genetics was the production of haploids through anther culture of Datura, achieved for the first time, again at Delhi University. Many countries, especially China, employed the technique of anther culture developed by the Indian scientists on a war-footing to improve the crop plants. The flowering of bamboo, which is a rare phenomenon, was demonstrated by NCL scientists in the 1980's. Development of triploid plants through endosperm culture was also first developed from Delhi University in the 1970's. Triploid plants produce seedless, juicy fruit, an example of which is juicy triploid watermelon.

Protocols have been developed for clonal multiplication of hundreds of plant species, which include trees, medicinal and aromatic plants, and endangered species from several laboratories across the Country. It is now quite apparent that India has made big strides in the realm of biotechnology. The galloping rate at which this advancement is continuing, holds out great promise for a better future for our country in the field of biotechnology.

Table 1: Milestones in the Field of Plant Tissue and Cell Culture

Year	*Important Discoveries and their references*
1838 & 1839	Cell theory, suggesting totipotency of cells. Schleiden M. J., *Arch. Anat., Physiol.* U. Wiss. Med. (J. Muller), 1838: 137-176; Schwann T., W. Engelman, No. 176 (1910).
1882	Plants synthesize organ-forming substances that are polarly distributed. Sachs J., *Arch. Bot. Inst.* Wurzburg, 2: 453 & 689. A.
1902	First but unsuccessful attempt of tissue culture using monocots. Haberlandt G., Sitzungsber Akad. Wiss. Wien, *Math.-Naturwiss.* Kl., 111: 69-92.
1904	First attempt in embryo culture of selected Crucifers. Hannig B., *Bot.* Zeitung, 62: 45-80.
1909	Fusion of plant protoplasts though the products failed to survive. Kuster E., *Ber. Dtsch. Bot. Ges.*, 27: 589-598.
1921	Cultivation of fragments of plant embryos. Molliard M., C. R. Soc. Biol. (Paris), 84: 770-772.
1922	Asymbiotic germination of orchid seeds. Knudson L., *Bot. Gaz.*, 73: 1-25. *In vitro* culture of root tips. Robbins W. J., *Bot. Gaz.*, 73: 376-390.
1924	Callus formation on carrot root explants by use of lactic acid. Blumenthal F. and Meyer P. Z. *Krebsforsch.* 21: 250-252.
1925	Embryo culture for interspecific crosses in Linum spp. Laibach F., Z. *Bot.*, 17: 417-459.
	Symbiotic germination of orchid seeds. Knudson L., *Bot. Gaz.*, 29: 345-379.
1929	Embryo culture to avoid cross incompatibility in Linum spp. Laibach F., *J Hered.*, 20: 201-208.
1934	*In vitro* culture of cambial tissues of different trees and shrubs failed. Guatheret R. J., *C. R. Acad. Sci. (Paris)*, 198: 2195-2196.
	Successful long-term culture of tomato roots. White P. R., *Plant Physiol.*, 9: 585-600.
	Identification of the first plant hormone, Indole Acetic Acid (IAA), leading to cell enlargement. Kogl F. *et al.*, Z. *Physiol. Chem.*, 228: 90-103.
1936	Embryo culture of different gymnosperms. LaRue C. R., *Bull. Torrey Bot. Club*, 63: 365-382.
1939	Successful in obtaining continuously growing cambial cultures of carrot and tobacco. Gautheret R. J., *C. R. Acad. Sci. (Paris)*, 208: 118-120; Nobecourt P., *C. R. Soc. Biol. (Paris)*, 130: 1270-1271; White P. R., Am. J. Bot., 26: 59-64.

1940 Culture of cambial tissue of *Ulmus* to study adventitious shoot formation. Gautheret R. J., *C. R. Acad.* Sci., 210: 632-634.

1941 Coconut Milk used for growth and development of very young *Datura* embryos. Overbeek J. van *et al.*, *Science*, 94: 350-351.

1942 Observation of secondary metabolites in plant callus cultures. Gautheret *R. J. Bull. Soc. Chim.* Biol. 41: 13.

1943 - 1950 Tumor-inducing principle of crown gall tumors identified. Braun *A. C. Phytopathol.* 33: 85-100 & P. N. A. S. USA 45: 932-938.

1944 First *in vitro* culture of tobacco, to study adventitious shoot formation. Skoog F., *Am. J. Bot.*, 31: 19-24.

1945 Cultivation of excised stem tips of Asparagus. Loo S. W., *Am. J. Bot.*, 32: 13-17.

1946 First whole plants of Lupinus and Tropaeolum from shoot tips. Ball E., *Am. J. Bot.*, 33: 301-318.

1948 Formation of adventitious shoots and roots in tobacco. Skoog F. and Tsui C., *Am. J. Bot.,* 355: 782-787.

1949 Culture of fruits *in vitro*. Nitsch J. P., *Science*, 110: 499.

1950 Organs regenerated from callus of Sequoia. Ball E., *Growth*, 14: 295-325.
First successful cultures of Monocots using coconut milk. Morel G *C. R. Acad. Sci.*, 230: 2318-2320.

1951 Culture of excised ovaries *in vitro*. Nitsch J. P., *Am. J. Bot.*, 38: 566-577.
Chemical control of growth and organ formation in culture demonstrated. Skoog F., *Annee Biol.*, 26: 545-562.

1952 Virus-free Dahlia through meristem culture. Morel G. and Martin C., *C. R. Hebd. Seances Acad. Sci.* (*Paris*), 235: 1324-1325.
First successful micro-grafts. Morel G and Martin C., *C. R. Acad. Sci.* (*Paris*), 235: 1324-1325.

1953 Haploid callus from pollen grain of *Ginkgo biloba*. Tulecke W. R., *Science*, 117: 599-600.

1954 First calli produced from a single cell by use of nurse cultures. Muir W. H. *et al.*, Science, 119: 877-878.

1955 Discovery, structure and synthesis of Kinetin. Miller C. *et al.*, *J. Am. Chem. Soc.*, 77: 1392 & 2662-2663.

1956 *In vitro* cultivation of normal and tumor tissues of *Picea glauca*. Reinert J. and White P. R., *Physiol. Plant.*, 9: 177-189.

1956 US patent No. 2747334 for: Production of substances from plant tissue culture of *Phaseolus* by Routien J. B. and Nickell L. G.

1957 Discovery that root or shoot formation in culture depends on auxin: cytokinin ratio. Skoog F. and Miller C. O., *In vitro Symp. Soc. Exp. Biol.*, No. 11: 118-131.
Culture of excised anthers of *Allium cepa*. Vasil I. K., *Phytomorph.*, 7: 138-149.

1958 *In vitro* culture of excised ovules of *Papaver somniferum*. Maheshwari N., Science, 127: 342.
Regeneration of somatic embryos from nucellus of Citrus ovules. Maheshwari P. and Rangaswamy N. S., *Ind. J. Hort.*, 15: 275-281.
Pro-embryo formation in callus clumps and cell suspension of carrot. Reinert J. and Steward F. C., *Naturwiss.*, 45: 344-345.

Growth and development in suspension cultures. Steward F. C. *et al.*, *Am. J. Bot.*, 45: 693-708.

1959 Production of large amounts (134 L) of plant tissue by submerged culture. Tulecke W. and Nickell L. G., *Science*, 130: 863-864.

1960 First test tube fertilization in *Papaver rhoeas*. Kanta K., *Nature*, 188: 683-684.

Enzymatic degradation of cell wall for protoplast formation. Cocking E. C., *Nature*, 187: 927-929.

Vegetative propagation of orchids by meristem culture. Morel G., *Am. Orchid Soc. Bull.*, 29: 495-497.

1962 Development of MS medium. Murashige T. and Skoog F., *Physiol. Plant.*, 15: 473-497.

In vitro flower induction in tobacco Aghion D., *C. R. Acad.* Sci., 255: 993-995.

1964 First haploid plants from *Datura* by androgenesis. Guha S. and Maheshwari S. C., *Nature*, 204: 497 and *Nature*, 212: 97-98 (1966).

Regeneration of roots and shoots on callus of *Populus tremuloides*. Mathes M. C., *Phyton*, 21: 137-141.

1965 Differentiation of tobacco plants from a single isolated cell in microculture. Vasil V. and Hildebrandt A. C., *Science*, 146: 76-77 & 150: 889-892.

Protocorm formation in orchids *in vitro*. Morel G., *Cymbidium Soc. News*, 20: 3

1967 Flower induction in *Lunaria annua* by vernalization *in vitro*. Pierik R. L. M., See Pierik R. L. M., (1987) *In vitro Culture of Higher Plants*. Martinus Nijhoff Publishers, Dordrecht.

1967 Yields of secondary products in cell culture equal to those of intact plants of *Ammi visnaga*. Kaul B. and Staba E. J., *Planta Med.*, 15: 145-156.

Haploid plants from pollen grains of tobacco. Bourgin J. P. and Nitsch J. P., *Ann. Physiol. Veg.*, 9: 377-382 & 10: 69-81.

1969 Protoplast isolation from suspension culture of *Hapopappus gracilis*. Ericksson T. and Jonassen K., *Planta*, 89: 85-89

1970 Selection of biochemical mutants in tobacco. Carlson P. S., *Science*, 168: 487-489.

Hybrid embryo culture and subsequent chromosome elimination for haploid production in Barley. Kasha K. J. and Kao K. N., *Nature*, 225: 874-875

Protoplast fusion. Power J. B. *et al.*, *Nature*, 225: 1016-1018.

1971 Plant regeneration from mesophyll protoplasts of tobacco. Takebe I. *Et al.*, *Naturewiss.*, 58: 318-320.

1972 Interspecific hybridization of *Nicotiana* spp. using protoplasts. Carlson P. S. *et al.*, P. N. A. S. (USA), 69: 2292-2294.

1973 Cytokinins found to be capable of breaking dormancy in Gerberas. Pierik R. L. M. *et al.*, Sci. Hort., 1: 117-119.

1974 Induction of branching by cytokinins in Gerbera shoot tips Murashige F. *et al.*, *Hortsci.*, 9: 175-180.

Regeneration of haploid Petunia plants from protoplasts Binding R. J., Z. *Pflanzenphysiol.*, 101: 119-130.

Fusion of haploid protoplasts to form polyploids. Melchers G. and Lalib G., *Mol. Gen. Genet.* 135: 277-294.

Bio-transformation in PTC. Reinhard E., In: *Tissue culture and Plant science*, Street H. E. (ed.), Acad Press NY,: 433-459.

Ti plasmid as the tumor inducing principle in crown gall. Zaenen I. *Et al., J. Molec. Biol.*, 86: 109-127; Larebeke N. van *et al., Nature*, 252: 169-170.

1975 Positive selection of maize callus culture resistant to *Helminthosporium maydis*. Gengenbach B. G. and Green C. E., *Crop Sci.*, 15: 645-649.

1976 Shoot induction from cryo-preserved shoot tips of carnation. Seibert M., *Science*, 191: 1178-1179.

Protoplast fusion of *Petunia hybrida* with *P. parodii*. Power J. B. *et al., Nature*, 263: 500-502.

Octopine and Nopaline synthesis and break-down is regulated by Ti plasmid. Bomhoff G. et al., *Molec. Gen. Genet.*, 145: 177-178.

1977 Successful integration of T-DNA in plants. Chilton M. D. *et al., Cell*, 11: 263-271.

Cultivation of tobacco cells in 20,000 L bioreactors. Noguchi M. *et al., Plant Tissue Culture & its Biotechnological Application*, Springer Verlag, Berlin,: 85-94.

Development of two-stage culture medium for suspension cell cultures. Zenk M. H. *et al., Plant Tissue Culture & its Biotechnological Application*. Springer Verlag, Berlin,: 27-43.

1978 Somatic hybridization of tomato and potato. Melchers G. *et al., Carlsburg Res. Comm.*, 43: 203-218.

Industrial scale fermentation of plant cells for production of shikonin. (Selection of cell lines with higher yield of secondary products). Tabata M. *et al., Frontiers of Plant Tissue Culture* 1978, Univ. Calgary Press, Calgary,: 213-222.

1979 Alginate beads used for plant cell immobilization for biotransformation and secondary metabolite production. Brodelius P. *et al., FEBS* Lett., 103: 93-97.

Co-cultivation procedure developed for the *Agrobacterium* mediated transformation of protoplasts. Marton L. *et al., Nature*, 277: 129-131.

1980 The use of immobilized cells for bio-transformation of digitoxin into digoxin. Alfermann A. W. *et al., Planta Medica*, 40: 218.

1981 Introduction of the term somaclonal variation. Larkin P. J. and Scowcroft W. R., *Theor. Appl. Gen.*, 60: 197-214.

Isolation of auxotrophs by cell colony screening in haploid protoplasts of *Nicotiana plumbaginifolia* treated with mutagens. Sidorov V. *et al., Nature*, 294: 87-88.

Use of a hollow fibre reactor for secondary metabolite production. Shuler M. L., *Ann. NY Acad. Sci.*, 369: 65-79.

1982 Naked DNA transformation of protoplasts. Krens F. A. *et al., Nature*, 296: 72-74.

Electrofusion of protoplasts Zimmermann U., *Biochim. Biophys. Acta*, 694: 227-277.

1983 Intergeneric cybrid in radish and rape. Pelletier G. *et al., Molec. Gen. Genet.*, 191: 244-250.

First industrial production of secondary metabolites by suspension cultures of *Lithospermum* spp. by Mitsui Petrochemicals.

Beneficial use of elicitors in cell suspension cultures. Wolters B. and Eilert U. *Dtsch. Apoth. Zeitg.*, 123: 659-667.

1983 Co-integrate type of vectors designed for *Agrobacterium* transformation. Zambryski P. *et al.*, *EMBO* J., 2: 2143-2150.

1984 Transformation of *Nicotiana* protoplasts with plasmid DNA and regeneration of transformed plants. Paszkowski J. *et al.*, *EMBO* J., 3: 2717-2722.

1985 Infection and transformation of leaf discs with *Agrobacterium tumefaciens* and regeneration of transformed plants. Horsch R. B. *et al.*, *Science*, 227: 1229-1231.

Development of disarmed Ti plasmid vector system for plant transformation. Fraley R. T. *et al.*, *Bio/Technol.*, 3: 629-635.

Development of binary vector system for plant transformation. An G. *et al.*, *EMBO* J., 4: 277-284.

Gene transfer in protoplasts of Dicot and Monocot plants by electroporation. Fromm M. E., *P. N. A. S.* (*USA*), 82: 5824-5828.

Hairy root production for the first time in *Hyoscyamus muticus*. These roots produced more hyoscyamine than *in planta*. Flores H. E. and Filner *P., Primary & Secondary Metabolism of Plant Cell Cultures*. Springer Verlag, (Eds. Neumann K. H., Barz W. and Reinhard E.): 174-186.

1986 Transformation of tobacco protoplasts by direct DNA microinjection. Crossway A. *et al., Mol. Gen. Genet.*, 202: 179-185.

1987 Use of microprojectile gun for particle bombardment for genetic transformation and recovery of individuals showing transient gene expression. Klein T. M. *et al., Nature*, 327: 70-73.

First monocot (Asparagus) transformation by *Agrobacterium tumefaciens*. Bytebier B. *et al., P. N. A. S.* (*USA*), 84: 5345-5349.

Microinjection for direct DNA delivery into plant cells. Miki B. L. A. *et al., Plant DNA Infectious Agents.*, vol. 3: 249-265.

1988 Recovery of stable transformants through particle bombardment. Klein T. M. *et al., P. N. A. S.* (*USA*), 85: 4305-4309.

Automated mass propagation with organogenesis and embryogenesis. Levi R. *et al., Biotechnol.*, 6: 1035.

1990 Plant transformation by microinjection of intact plant cells. Neuhaus G., *Physiol. Plant.*, 79: 213-217.

Electroporation of intact plant tissues for direct DNA delivery. Dekeyser R. A. *et al., Plant Cell*, 2: 591-602.

Silicon carbide fiber-mediated DNA delivery in plant cells. Kaeppler H. F. *et al., Plant Cell Rep.*, 9: 415-418.

1991 Cryopreservation of alkaloid-producing cell culture of *Catharanthus*. The cells retain the property of alkaloid synthesis even after thawing. Lynch P. T. and Benson E. E., *Rice Genetics II*, IRRI, Manila, Philippines,: 321.

Production of first transgenic plants of a conifer (*Larix decidua*, by

	Agrobacterium rhizogenes mediated transformation). Huang Y. *et al.*, *In vitro Cell Dev. Biol.*, 27: 201-207.
1992	Successful metabolic engineering of *Atropa belladona* for increased alkaloid production. Yun D.-J. *et al., P. N. A. S.* (*USA*), 89: 11799-11803.
	Herbicide resistant rice plants through PEG mediated transformation of protoplasts. Dutta S. K. *et al.*, *Plant Mol. Biol.*, 20: 619-629.
1993	*In vitro* fertilization with isolated single gametes resulting in zygotic embryogenesis and recovery of fertile maize plants. Kranz E. and Lorz H., *The Plant Cell*, 5: 739-746.
	"Green Hairy roots" showing photoautotrophy due to development of photosynthetic ability. Flores H. E. *et al., Plant Physiol.*, 101: 363-371.
1994	Marketing of genetically engineered tomato: '*Flavr-Savr*'
	Improved plant transformation through electrophoresis. Giesbach R. J., *Plant Sci.*, 102: 81-89.
1996	Development of 'agrolistic' method of plant transformation. Hansen G. and Chilton M. D., *P. N. A. S.* (*USA*), 93: 14978-14983.
	Development of a binary bacterial artificial chromosome (BBIC) vector for Agrobacterium-mediated transformation (Transfer capacity of 150 kb). Hamilton C. M. *et al., P. N. A. S.* (*USA*), 93: 9975-9979.

2
Laboratory Design and Development

Plant tissue culture laboratories have design requirements that distinguish them from other types of laboratories and some needs are unique for a particular business for which the laboratory is set up. Therefore, careful initial planning is desirable in order to avoid problems related to special requirements.

2.1. THE BUILDING

2.1.1. Structures and Utilities

The selection of interior building materials should be made relative to durability, availability and cost. A single level structure that provides easy access to the different work areas is preferable for safety reasons and because typical tissue culture activities require frequent movement of materials between areas.

A tissue culture laboratory needs a source of good quality water, although some variation in quality of incoming water can be tolerated because it can be purified to different degrees in the laboratory. The details of water access and disposal need thorough consideration in a location where public water and sewer facilities are not available.

Even where a reliable source of electricity is available, power outages can occur, particularly in areas prone to thunderstorms, ice storms and high winds, gutsy winds and heavy rains. Rains in some developing countries automatically forces electricity departments to cut off the supply for safety reasons. One such case is in the city of Bangalore, India. A power failure can be critical for temperature control of cultures in the growth room and the cold storage. Power failure could be suicidal for plants in the green house kept for hardening. Even an undue increase in temperature for a short duration can ruin plants kept for hardening. This is not only fatal at University research situations and also can result in substantial or colossal losses at the industrial level. New companies set up are smart enough to have generators. At the University level, it is advisable to have a generator to provide electricity in an emergency as a standby. The electric wiring for these critical areas must be planned initially in such a way that the heating/cooling and lights can be switched over readily to generate power. Genera-tors are available that start up automatically if the main power source fails.

2.1.2. Layout

The types of activities to be performed and their requirements in terms of cleanliness, equipment and pattern of movement have led to the designation of areas that are common to most plant tissue culture laboratories: media preparation room, washing/sterilizing room, transfer room, growth rooms, office, general storage, culture removal, planting and/or shipping. The layout of these rooms, their pass-through windows, doors and hallways must allow a work flow pattern that maintains maximum cleanliness and promotes minimal backtracking.

The office, general storage and shipping areas can be maintained under normal sanitation conditions, but it is desirable to isolate the media preparation, transfer, growth and cold storage rooms into a 'clean zone' with controlled access, perhaps across a sticky mat to remove debris that could be tracked into the area. Some laboratories require hair, clothing and shoe covers or an entire clothing change for anyone entering the restricted area. The degree of sanitation is related to the external environ-ment and the amount of traffic expected to move into the clean area.

2.2. THE ROOMS

A plant tissue culture laboratory has several functional areas, whether it is designed for teaching or research and no matter what its size or how elaborate it is (Fig. 1). The different areas of a plant tissue culture laboratory includes:

1. Media preparation area
2. Washing/Sterilizing area
3. Inoculation room/Sterile transfer room
4. Culture room/Incubation room/Growth room
5. General storage area
6. Research area
7. Planting and/or Shipping area

In some laboratories, media preparation area, washing and sterilizing area and storage area do come under an area called preparation area. Hence, basically, the laboratory could be divided into three or four areas i.e., preparation area, culture transfer area and growth room. Alternatively, it could be media preparation area, autoclaving area, culture transfer area and culture growth room. In some laboratories, like the laboratory of Unit for Advanced Propagation Systems,

Wye College, (now Imperial College), University of London, where the first author was pursuing Doctorate Studies, a room called dirty room was built attached to the laboratory but away from the sterile activities. It makes sense to have this area to clean the plant material brought from the field, which probably have higher amount of dirt and of course, the organisms attached to the plant material. In this area, a large sink, taps which carry water at a force, shelves, dust pans, brushes and all that is needed for general pre-cleaning of the plant material to be done should be provided.

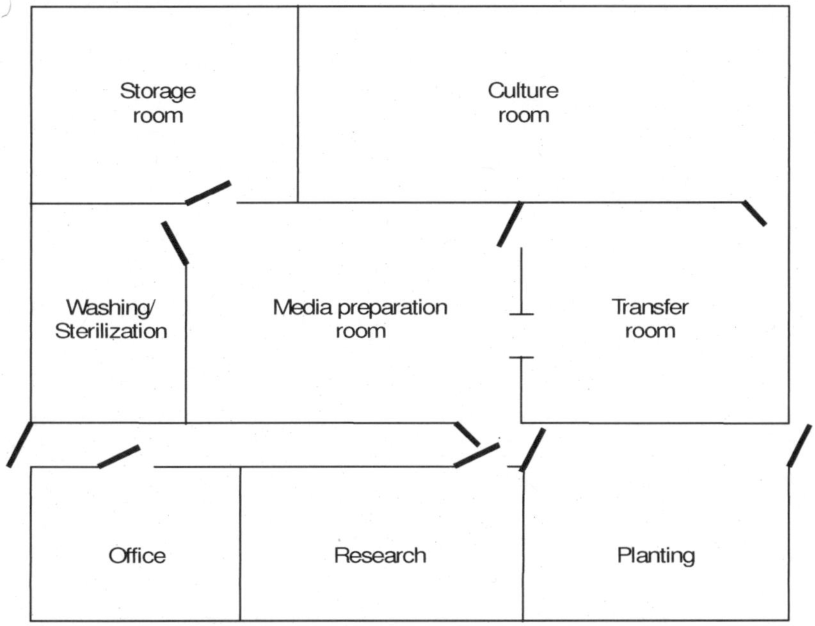

Fig. 1: Floor plan of a small tissue culture laboratory

2.2.1. Media Preparation Area

This is the central section where most of the activities are performed including media preparation as well as other activities that relate to handling of the tissue culture materials. The preparation area houses autoclaves and other sterilizing equipments such as ovens. Sources of distilled water as well as lines of electricity are provided. Preparation of field grown and green house plant materials for tissue culture initiation are best performed in a cleaning area partitioned from the main preparation section. This also serves as an area for the removal of discarded and contaminated cultures. Sinks and soaking tubs are provided for the purpose. Apart from the autoclaves and the oven, other equipments present in this area are magnetic stirrer, digital pH meter, electronic balance, etc.

This area is provided with floors and walls with a smooth surface to minimize dust accumulation and to facilitate cleaning. Windows may be permitted in this room, but should be sealed against external air. The number of doors to the room should be kept to a minimum within the constraints of local fire safety regulations or building codes. For phytosanitary reasons, the doors should open into a common hallway but not directly outdoors. An in-the-wall sliding door provides an alternative with minimal impact on usable wall space. If media preparation shares a room with washing/sterilizing, venting the steam that escapes while opening the autoclave aids in reduction of the heat load. Lighting in the area should be generous to promote easy visibility of work and employee productivity. Facilities for both tap and purified water are needed for the media preparation and the washing/sterilizing rooms.

Certain safety items desirable for the media preparation room include a general purpose fire extinguisher, fire blanket, an eye wash station, and a first aid kit.

2.2.2. Washing/Sterilizing Area

This room may be open into the media preparation room since fresh medium must be brought here for sterilizing and clean glassware must be returned to the preparation room for filling. An additional door opening into the hallway allows movement to and from the room without passing through the media preparation room. The room needs moisture tolerant floors and walls finished to withstand steam and cleaning. Although heat and steam from an autoclave and glassware washer are intermittent, an exhaust fan is needed to remove the warm moist air as and when necessary. In addition to a glassware washer and one or more autoclaves, the room needs to have a large sink, some counter top space, cabinets and mobile carts. If the laboratory is large enough to need multiple sinks, maximum flexibility is afforded by having one very long basin with multiple faucets and drains that can be subdivided as needed with plastic pans.

There are no special lighting needs in this room, but a brightly lit room is more pleasant to work in than a dimly lit one. Counter tops may be provided at a height for standing work. Tall mobile carts provide flexible holding space for items awaiting sterilization or washing and for clean jars. Thick floor mats are available that can be used in front of the cabinets and sinks. These reduce glass breakage and leg fatigue for operators working there.

2.2.3. Inoculation Room/Sterile Transfer Room

All the activities of sterile transfers are performed in this section (Plate 1). The area houses the laminar flow hoods needed to ensure the sterility required for the transfer operations. Temperature control is essential as heat is produced continuously from the flames of burners or electric sterilizers used in the hoods.

Since the transfer room is the site for sterile manipulations of plant tissue, cleanliness here is paramount. The room is constructed to minimize the collection of dust and for easy cleaning. Semi- or high-gloss finish on the walls and linoleum floors that curve up the walls a few centimetre will be easy to clean. There is no need for windows, but also no reason not to have them as long as they are tightly secured against external air. The number of doors to the room may be dictated by local safety codes but none should open directly outside unless required for safety. A sink is desirable for hand washing and rinsing disinfectant solution splashes. Use of mobile units permits flexibility in rearranging the room to accommodate

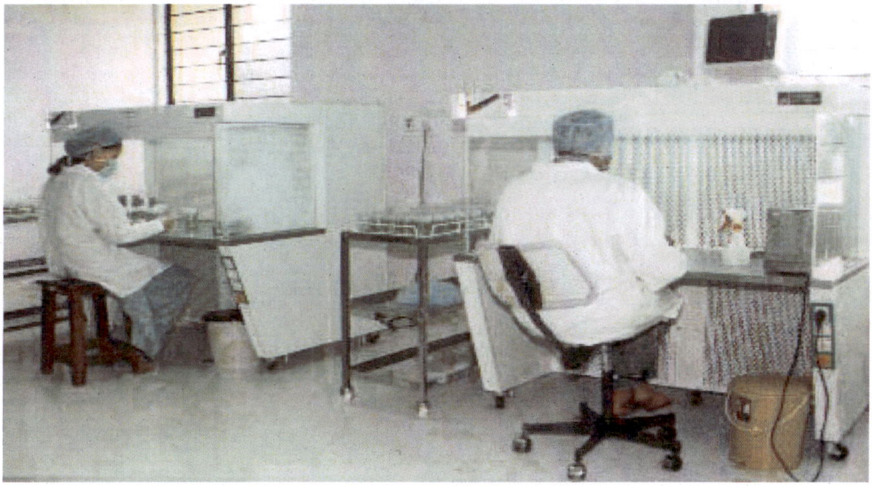

Plate 1: Plant transfer area

future changes. Standard laboratory push carts with solid shelves are valuable for transporting materials to and from each transfer station and for holding cultures to be transferred, fresh medium, used jars etc.

There should be direct access to the culture rooms or growth rooms for easy transfer of the inoculated culture vessels.

2.2.4. Culture Room/Incubation Room/Growth Room

The culturing or incubation facilities consist of individual rooms for housing units of culture racks (Plate 2). It is desirable to have more than one growth room to provide different temperatures or light cycles. There is minimal need for doors and access to the room should be limited to reduce the chance of introducing contaminants. Exterior windows are not desirable because external light cycle could interfere with that maintained in the room and temperature control would be more difficult.

Plate 2: A broader view of a part of Plant Tissue Culture Laboratory, showing spacious growth rooms

The culture room is maintained at a controlled temperature of 25 ± 2°C with the aid of air-conditioners. Uniformity of temperature throughout the room is important since 'hot or cold pockets' could affect culture survival and growth. While ceiling lights are desirable for working in the growth room during the dark cycle, the primary light sources are mounted on the shelves. The culture racks/ shelves are provided with separate sets of fluorescent tubes (Plate 3). Provision

Plate 3: Culture bottles arranged on racks in a culture room

is also there for variable photoperiod control in the culture rooms where diurnal control of illumination is achieved by using automatic timers (Plate 4).

There are multiple approaches to shelving in the growth room. Shelving units can be either stationary or mobile. A given space can accommodate more mobile units than fixed ones since mobile units can be pulled into the aisles between rows for access. The culture flasks, jars and petridishes can

Plate 4: Separate control chamber for growth room with automatic timers and other controls.

be placed on trays of suitable sizes on the shelves while metallic wire racks are used to hold the culture tubes (Plates 3 and 5). Each rack is labelled with the details of the experiment like name of the plant, explant, medium, date of culture, etc.

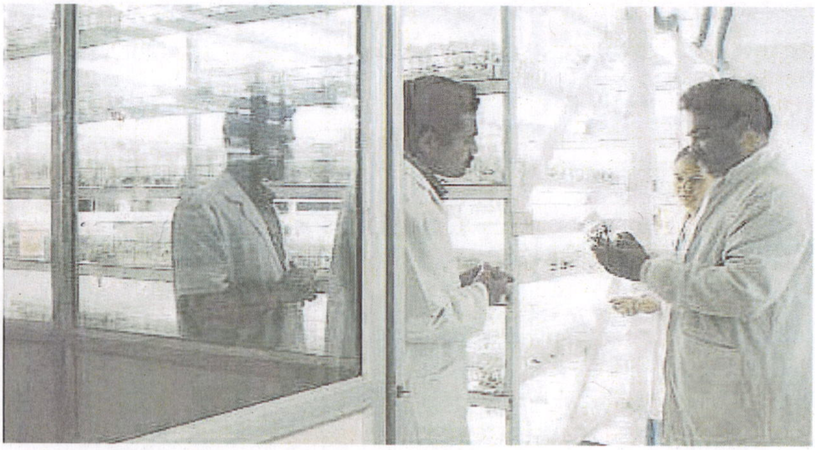

Plate 5: Growth chamber of Plant Tissue Culture Laboratory

It is advisable to connect the culture rooms with power generator for emergency power supply to maintain both light and temperature in the culture room (Plates 4 and 5).

2.2.5. General Storage Area

There are no special requirements for the floor and walls in the general storage room. Generous lighting and adjustable shelving would contribute to the ease of use.

The storage room consists of cabinets or shelves for storing the chemicals required for media preparation, explant treatment etc. As a practical and convenient approach, the chemicals are arranged in alphabetical order and assigned to specific areas. The growth regulators and such other chemicals are stored in the refrigerator.

2.2.6. Research Area

This area could house equipments and activities related to microscopic examination of plant material whenever required. Histological and histochemical, and biochemical analytical activities, assessing of degree and character of infections could be taken up. Some commercial laboratories do have a separate wing called Research and Development (R&D). This wing is precisely involved in developing production protocols and of course, taking up other required analytical works. For tissue culture, it is advisable to have inverted microscopes, which facilitates observation of cultures, which are kept in the conical flasks, bottles and other containers. For research laboratories at university level, it would be wiser to have microscopes with fluorescence and phase contrast facilities. Some microscopes are fitted with eyepieces in such a way that both tutor and a student could view the focused material at the same time. It would be appropriate to focus what is seen in the microscope on to a monitor or a large projection screen to aid teaching a group of students. Fluorescence microscopy aids examining tissues at a quicker pace. The tissues may not need to be subjected to treatments of longer durations, as is the procedure with conventional microscopy techniques. A microscope with a provision of epi-fluorescence would be sufficient to examine the plant tissue sections. Hence, the thickness of sections does not really matter. Sections taken from a razor blade is all that which is needed. Hence, use of microtomy, which probably needs good amount of skill could be avoided. The first author was able to handle hundreds of samples of sections of microshoots in a day with fluorescence microscopy technique.

2.2.7. Planting and/or Shipping Area

This area would be the logical place for initial plant cleanup, removing cultures for shipment and planting for the greenhouse, and for storing janitorial supplies. The culture handling would require good lighting, working tables and chairs and a sink plus appropriate materials for planting and shipping.

Improper handling of tissue cultured plants during hardening and planting *in vivo* leads to substantial loss of plants primarily due to dehydration. Some species of plants are very sensitive to low humidity conditions under *in vivo* situations. Exposing tissue cultured material of such species to a low humidity and higher temperature and light conditions even for a short duration would result in irreversible damage and death of plants. The first author has observed that exposure of *in vitro* raised Jackfruit plantlets to harsh conditions *in vivo* even for 30 seconds was enough to cause an irreversible damage. In contrast, plants such as *Anthurium* could survive harsh conditions for sufficiently a long time. However, it is important that all tissue cultured plants when handling under *in vivo* conditions soon after their removal from *in vitro* containers should be provided a high humidity condition. It makes sense to carry out the operation of transfer of *in vitro* plants to net pots or other containers in areas or rooms provided with air-conditioners or alternatively fogging facilities. It is also important that all the plantlets be treated with fungicides before transfer to *in vivo* conditions.

It is a practice to ship prehardened plants as such in the containers in which they are being raised. Hence, tissue cultured plants are raised in bigger containers such as conical flasks or boxes. In these containers, many number of propagules or plants could be accommodated. It is also a good practice to use polypropylene boxes or containers than glass containers to minimize damages during shipping. It is expected that the importing agency will have their own hardening units. These practices would ensure reduction of the cost of production as well as availability of plants to the customers at a cheaper rate. Now a days, some laboratories are finding ways of reducing the cost of production. One such way is producing plants in polypropylene packets instead of either bottles or other containers. Owing to reduction in the weight of the material, substantial savings in freight charges and also the polypropylene packets could be either dispensed with or if needed, could be reused.

In the commercial units or laboratories where no separate provision been made for handling during hardening, the operation of transfer of plants could be taken up in the hardening greenhouses itself. Alternatively, at least a shade house could be built for the purpose.

3

INSTRUMENTATION

The plant tissue culture laboratory houses all the necessary equipments needed for postgraduate teaching as well as research. The equipments, to name a few, are Laminar flow hoods, Horizontal and Vertical autoclaves, Incubator cum Centrifuge, BOD incubator, Growth chamber, Quartz distillation unit, Inverted microscope, etc.

3.1. LAMINAR FLOW HOODS

The sterile transfer room is provided with laminar flow hoods. The laminar airflow cabinet has a number of air blowers, which blow air through a high efficiency particulate air (HEPA) filter. The HEPA filter removes particles larger than 0.3μm. The resulting ultra clean air, which is free from fungal and bacterial contaminants flows at a velocity of about 27±3m/min across the width and height of the working area without turbulence thereby maintaining a continuous air current, which prevents the air from outside to enter the laminar area and thus keeps the cabinet without the risk of contamination.

There are two kinds of laminar flow hoods called as horizontal and vertical. For plant tissue culture work, use of horizontal flow hoods is advised (Plate 6). In a vertical hood, the filtered air blows down from the top of the cabinet while in a horizontal hood, the filtered air blows out at the operator in a horizontal fashion. They should never be used for bacterial or fungal work. Vertical laminar flow hoods, also known as biology safety cabinets are normally used for microbiological and clinical work. It is found best for working with hazardous organisms since the aerosols that are generated in the hood are filtered out before they are released into the surrounding environment. Horizontal laminar flow hoods are designed with horizontal airflows from back to the front. Since the air flows directly at the operator, horizontal hoods are not advisable for working with hazardous organisms, but are the best protection for your cultures. Both the chambers are provided with UV sterilizing tube and fluorescent tubes. The short wave UV light can be turned on for a few minutes to sterilize the surfaces of the hood, but we should be aware that only exposed surfaces would be accessible to the UV light. The switches to operate them are present outside the chamber so that the UV light can be safely switched on and off. UV sterilizing tubes are not absolutely needed. Without this, efficient working methodology could be worked out. Before initiating the culture operations, thorough swabbing of the laminar surfaces and the other surfaces of the laminar hood should be done. Electrical outlets are also provided for using steribead sterilizers and electrical balances. The working surface is made of stainless steel and is most durable and easiest to keep clean.

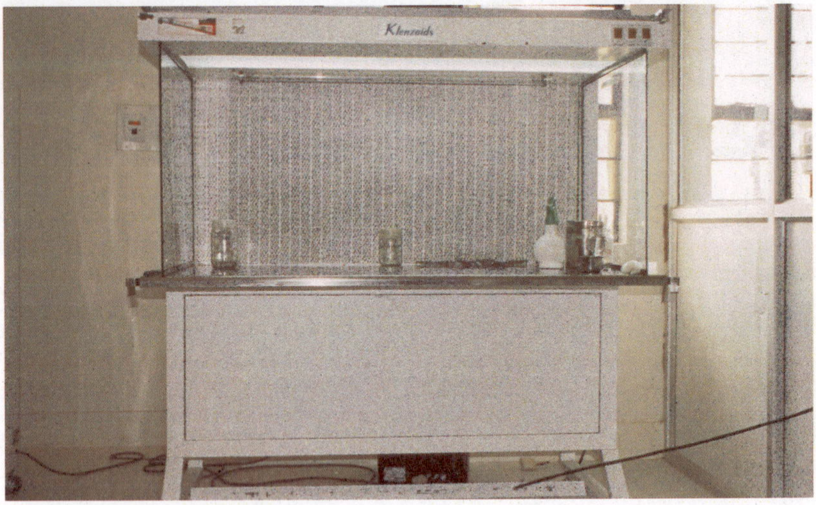

Plate 6: Horizontal Laminar Flow Hood

3.1.1. Operation and Management

- Wash hands with soap and water before starting any work.
- Wipe down the surface and sides of the laminar flow hood with 70% alcohol before starting the work.
- Set up all the materials including the containers, media and other utensils within the chamber and turn on the UV lamp. For achieving a satisfactory degree of aseptic environment inside the cabinet, a 20-25min exposure to UV light is enough. The forceps and scalpels are sterilized by keeping them in the steribead sterilizer for 15min.
- Before starting the work, switch off the UV lamp and switch on the fluorescent tubes and airflow. While working, be careful not to disturb the airflow patterns that could otherwise introduce non-sterile air.
- Cover the culture containers as quickly as possible after an operational step is completed.
- When finished, turn the sterilizer off, remove all the materials, and wipe the surface clean with alcohol.

3.1.2. Caution while using Laminar Flow Cabinets, Particularly Bunsen Burners and UV Lights

Utmost care should be taken while using UV for culture operations. In laboratories wherever UV lights are operated, they are normally switched on for a particular time before carrying out the actual culture work. It is essential that these lights must be switched off before starting the work. When the fluorescent light is switched on, it is not uncommon to forget switching off the UV light as this gets merged with the normal fluorescent light. Any such lapses of keeping the UV on while working even for a short duration of 10 to 15 minutes would lead to serious skin damage leading to severe irritation. Eyes, being a very sensitive part, is seriously affected, and sometimes leading to permanent damage. Hence, it is advisable to ignore the use of UV light for culture transfer operations. It is also a common practice especially among students (as is normally observed in our laboratory) to switch on UV for longer durations and sometimes unscrupulously leaving it on for hours. This not only could pollute the environment in the laboratory but also reduce the shelf life of bulb. UV bulbs have a definite life span in terms of its efficacy to cause sterilization.

Precautions should also be taken while using Bunsen burners in the laminar flow cabinets. Flame sterilization of the forceps followed by its re-dipping into alcohol can lead to major fire accidents. The first author has experienced and also has witnessed two other students when he was pursuing doctorate studies. If there were no fire extinguishing devices, the results could have been catastrophic. It is also observed at the Laboratory of Plant Tissue Culture headed by the first author that the students tend to use Bunsen burner even though steribead sterilizers are available. This is because they believe that sterilization is best done with a Bunsen burner. The fact is sterilization could be accomplished best with the steribead sterilizers too. Keeping faith with the steribead sterilizers is certainly a wiser option than naked flame sterilization. Nowadays, there are other kinds of sterilizers which are safer and of course, does not work on the naked flame devices. If steribead sterilizers are maintained in good condition, it augurs well for both research as well as commercial activities. An important maintenance procedure to be followed is with regard to cleaning the steribeads, which would have turned brown in colour due to regular usage. The steribeads have to be cleaned by soaking them in a solution of chromic acid plus sulphuric acid (diluted). In addition, there may be several inflammable materials like cotton in the vicinity of the burner and there is every chance for it to catch fire if proper care is not taken. Stressful or working under disturbed mindset or not being composed while working in the laminar flow cabinet especially with Bunsen burner is always dangerous. Naked flames can always cause injury to the skin. Hence, it is always advisable to go for a steribead sterilizer for the sterilization of the instruments used instead of flame sterilization with the aid of Bunsen burners.

3.2. AUTOCLAVES

Steam sterilization of materials is a dependable procedure for the destruction of all forms of microbial life. Autoclave is an instrument in which, saturated steam under pressure effects sterilization. An autoclave is basically a double walled metallic vessel made of thick stainless steel or copper, one end of which has an opening fitted with a tightly closing door. The opening is meant for keeping the materials to be sterilized.

Electrically operated vertical and horizontal autoclaves (Plate 7), horizontal autoclaves, provided with safety valve, pressure gauge and steam release cock, are used for the sterilization procedures of the glasswares and media. There are also valves to hold the steam and another to send it into the sterilizing chamber. There is also an exhaust valve below to let the steam escape from the bottom of the inner chamber. All the glasswares and media are sterilized by autoclaving them at a temperature of 121°C at 105kPa for 20 minutes. The contaminated cultures also have to be autoclaved before the contents are disposed off.

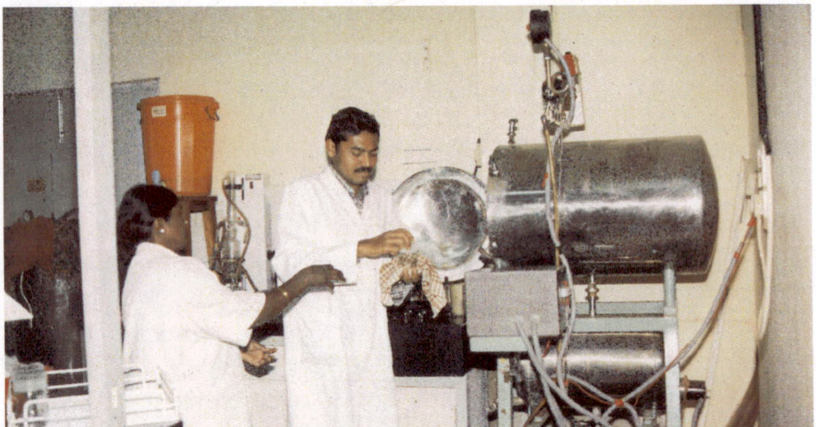

Plate 7: Culture bottles being placed in a Horizontal autoclave

The working principle of an autoclave is similar to that of the domestic pressure cooker, wherein sterilization is achieved by steam under high pressure.

Successful components of sterilization include validation of decontamination effectiveness, which includes monitoring of temperature, pressure and cycle duration.

Temperature: Heating is achieved by means of saturated steam under high pressure to attain a chamber temperature of at least 121°C (250°F).

Time: The time is measured after the temperature of the material being sterilized reaches 121°C. The material is kept at that temperature for a minimum of 20 minutes.

Contact: Steam must come in contact with all areas of the load. Autoclave bags should be left partially open during decontamination to allow steam to penetrate into the bag. Air pockets or inadequate steam supply will cause sterilization failure.

3.2.1. Operation and Management

- Before sterilization, care should be taken to see that the tank is filled with water to the level as indicated. Negligence in terms of not filling water can render the coil useless and it can also spoil the electrical circuit thus may be a cause for electrical shock too.
- While loading the bottles or the tubes into the autoclave, ensure that the caps are not completely screwed. Sometimes, extreme build up of pressure can blow off the lid especially when the screws are completely tightened. Under other circumstances such as with overused caps, screwing the caps may not be hazardous, as they tend to become loose by themselves under pressure. The media/glasswares to be sterilized are then kept in the autoclave and the lid is closed tightly.
- When the steam starts coming out of the valve, it should be closed. Now the pressure starts building inside. Sufficient ventilation should be ensured in the room where autoclaves are kept as this takes care of reducing the vapours in the area. Mind you, a higher water vapour in the laboratory, unchecked, may contribute towards increased microbial contamination.

Note:
1. All plastics can not be autoclaved. Polypropylene and polycarbonate will be stable but polyethylene will not. The different types of plastics can be identified by looking for the initials imprinted on the bottom of the containers (PP- Polypropylene, PC- Polycarbonate, PE- Polyethylene, HDPE- High Density Polyethylene).
2. To prevent the bottoms of bottles from breaking, place them in a tub with 1-2 inches of water.
3. Autoclaving new glasswares for 90 minutes may partially temper them, increasing their strength.

3.2.2. Caution

While working with the autoclaves, it is very important to be composed and not do things in a hurry. It is our experience that the students tend to open the lid immediately after the autoclaving time when the pressure is still not completely released. This sort of handling could be life threatening. Hence, utmost care and attention is needed when operating the autoclave.

3.3. DISTILLATION UNIT

It consists of a heating mantle for round bottom flask with heating surface made of glass with nichrome wire. Single or double distillation assembly can be used for obtaining single or double distilled water.

3.3.1. Reverse Osmosis Water

Reverse osmosis has long been considered the most comprehensive solution to obtain extra pure water for laboratories. When two volumes of water (or other solvent) are separated by a semipermeable membrane, water will flow from the side of low solute concentration to the side of high solute concentration. The flow may be stopped, or even reversed by applying external pressure on the side of high solute concentration. In such a case the phenomenon is called reverse osmosis.

The equipment that supplies reverse osmosis water consists of a four stage system. The first stage consists of a sediment/carbon pre-filter (5 microns), where particulate matter and chlorine are removed. The second stage is the heart of the system where, the water is forced against a thin film composite membrane which traps the impurities through reverse osmosis. At the third stage activated carbon aerates and freshens the water and sends it into a space-saving 2.1 gal storage tank. At the final stage a carbon post filter "polishes" the water to get ultra-filtered water.

3.3.2. The Millipore System

The Millipore systems are designed to produce ultrapure water. The system comprises a pre-filter unit, a Milli-Ro unit, a storage tank for water produced from the Milli-Ro unit, and a Milli-Q unit. The Milli-Ro unit utilises the Reverse Osmosis process to produce laboratory grade water, which is stored in the reservoir. The reservoir tank is protected against contamination from air-borne bacteria by a bacterial retentive vent filter. The water from the reservoir tank is then passed to a Milli-Q unit, which produces ultrapure water on demand.

Compact, battery-powered systems are also available for users requiring low volumes of pure water (3-5l/day). Such plug-and-use purification packs are tailored to the feedwater source and can be fitted to the tap to provide deionized water directly. Such a device is economical, easy to use, requires less maintenance and is suited for small laboratories.

In a laboratory, especially of universities and R & D units of companies, it is wiser to supply distilled water throughout the length and breadth of the laboratory. This ensures smooth and efficient functioning of the laboratory, thus also saves time with the personnel who may have to move to areas at some distance to obtain water. Designing of twin water basins for cleaning purposes is a must. The culture bottles to be used may be washed in normal tap water and then given a rinse in single and subsequently with double distilled water. Alternatively, the culture bottles which are contaminated should be autoclaved, cleaned taking care to remove the agar and then washed with detergent or laboratory cleaning solutions. Subsequently, they have to be rinsed in single and double distilled water. Later, these are stored in a dust free environment till reused.

3.3.3. Operation and Management

- There should be provision for uninterrupted water supply.
- There should be free flow of water through the condenser.
- Once the apparatus is put on, distilled water can be collected in drops in a container.

In the laboratories, if there is only a double distillation apparatus, especially when the quality of water is hard, care should be taken to supply single distilled or deionized water only to supply to the double distillation unit. Directly supplying tap water to the double distillation unit will result in the accumulation of scales inside the glass tubings. This would slow down the distillation to a considerable extent thereby costing in time as well as increased cost owing to decreased usage of electricity per unit water harvest.

3.4. DIGITAL pH METER

A digital pH meter with combined pH electrode in single rod assembly is provided to adjust the pH of the culture media. The electrode, when dipped in the solution, draws hydrogen ions towards its outer surface. The number of hydrogen ions accumulating on the membrane depends on their concentration in the external solution.

Correct pH of the medium (usually between 5.6-5.8) is important as highly alkaline or highly acidic pH values affect the nutrient uptake by the cultured tissues. The instrument must be calibrated against buffers of known pH before measuring the pH of the specimen.

3.4.1. Operation and Management

- Wash the electrode of the pH meter with distilled water.
- Before using the pH meter for measuring pH, it is advisable to calibrate with different pH buffers. Take 10-15ml of standard buffer of pH 4.0 in a 25ml beaker and dip the electrode in this standard buffer. Adjust the pH to 4.0 by operating the designated knob.
- Wash the electrode and check with another buffer solution of pH 7.0. Repeat the adjustment.
- Dip the electrode in the medium and read the pH of the medium. If the pH is less than 5.5, add 1N NaOH dropwise and if it is more than 5.8, add 1N HCl to adjust the pH to 5.6-5.8.
- After reading the pH, wash the electrode thoroughly with distilled water.

- Since the electrode is very delicate, it should not come in direct contact with the walls of the beaker.
- When not in use, ensure that the bulb of the electrode is dipped in distilled water/KCl solution.

3.5. ORBITAL SHAKER

Orbital shaker is mainly used for the maintenance of suspension cultures, which require constant agitation for adequate aeration. It also facilitates dispersion of cells.

The instrument consists of a platform and a speed regulator. The platform of the shaker is fitted with clips of appropriate sizes for holding the flasks. The platform/disc can be rotated at low speed, thereby the cells and tissues in the flask are alternately bathed in the culture medium and thereby exposed to both nutrient and air. The speed regulator connected to the shaker is used to adjust the revolution per minute as per the requirement.

3.5.1. Operation and Management

- Place the flasks securely within the clips.
- Switch on the main power supply and push the lever of the instrument to 'On' position.
- The speed control knob is then turned in anticlockwise direction completely so that digital display reads zero.
- The knob is then turned clockwise until the display reads the required rpm.

3.6. INVERTED MICROSCOPE

Microscopes with an inverted-style frame are designed primarily for tissue culture applications (Plate 8). The light source and condenser are present

Plate 8: Inverted microscope

at the top above the stage pointing down and the objectives and turret are below the stage pointing up. As a result, one is looking up through the bottom of whatever is holding the specimen and is sitting on the stage rather than looking at the specimen from the top, typically through a cover glass, as on a conventional microscope. Because of its configuration, an entire culture or large sample in a relatively large container such as a Petri dish can be placed on the stage and it is possible to look at the entire contents of the container under more natural and less stressed conditions. The cover of the container slows evaporation greatly. Further information on microscopy is given in Chapter 29 on microscopic and photographic techniques in plant tissue culture.

3.7. OTHER INSTRUMENTS

1. Hot air oven
2. Magnetic stirrer
3. Heater

3.8. CULTURE VESSELS

Different types of vessels have been used to culture plant materials. While in some cases, the choice of culture vial is dictated by the nature of the experiment, in others, it has been guided mainly by the convenience and fancy of the worker. For standard tissue and organ culture work, glass test tubes have been widely employed. Wide mouthed glass bottles of different sizes are also used. Generally, only borosilicate or pyrex glasswares are used. The glass culture vials and other labware required for media preparation may be replaced by suitable plasticware, some of which are autoclavable. For culture tubes and flasks, traditionally, cotton plugs sometimes wrapped in cheese clothes have been in use. However, alternative closures can be used like autoclavable plastic-caps for test tubes.

For experimental purposes, the choice of glasswares has to be carefully made. It has been reported by scientists that the responses, though not highly significant, vary depending on the culture vessels used. These differences may probably be due to variations in the amount of gases in the container and more particularly, the amount of light passed through the sides of the vessels. While selecting the culture bottles or tubes, it is important that they are of high quality. Badly finished bottles or culture vessels account for low responses of cultures, results in breakages, hence accounting for loss of plant material as well as culture vessels. The bad quality glass also comes in the way of taking better pictures (photographs) as the reflections of light from the surface of the glass of bottles or tube hamper the quality of pictures. The caps selected for the bottles should be of polypropylene capable of withstanding high temperature and steam. The threads of the caps should be precise enough to match that on the bottleneck. The caps selected for the tubes are normally of push types. It is important here also that the quality of caps is of high standard. In the absence of threads, it is important that these caps withstand high temperature and steam, as any little expansion would result in loose fitting with the tube. Continued autoclaving of these caps normally render them useless, as they do not fit the mouth of the tubes any more. The caps for the bottles and the tubes selected should be of translucent ones. In most of the laboratories, the tubes are fitted on the culture racks in such a way so as to focus the light over the culture vessels. Under such circumstances, the opaque covers used will certainly reduce the amount of light falling on the cultures. Nowadays, in many laboratories, and in our laboratory too, it is a practice to cover the gap between the screwed cap and the bottle with polypropylene strips or cling film sheets. This is especially done to reduce contamination. The contamination, which may be carried through insect pests such as dust mites or ants, could be prevented as the gap is covered by the polypropylene. Any contamination which may be carried through the air in cases of culture bottles where the cap is faulty or the cap is not tightened properly, polypropylene covering may be avoided. Parafilm strips could also be used similar to the usage of cling films. The advantage of parafilm is that it is stretchable. This quality also accounts for a disadvantage. If the temperature in the growth room is higher or increased heat accumulation results in the expansion of the parafilm and becomes loose, thus does not serve the purpose for which it was intended.

It is important and makes sense to use the right kind of culture vessels, be it a tube, a bottle or a conical flask. Though normally in laboratories, the culture vessel selected depends on the availability of such vessels in the laboratory. It is wise to minimize the use of media and hence a smaller bottle or a small sized tube should be used if the plants selected grow only to a limited height. We have seen in our laboratory, students tend to use a tube which is available for example; a tube of 150mm in length containing 30ml of media is actually a waste to grow a callus from a particular tissue. It is wiser to use a 100mm tube for this purpose. As regards a bottle, it is better to use a baby jar in most cases. A baby jar available in India has a volume of nearly 300ml. It is advisable and advantageous to use a bottle even smaller probably which can hold 150 to 200ml. College laboratories can look for smaller bottles available in the markets. These are normally the bottles made for storing food materials (processed products). Bigger or commercial laboratories, if their requirement is high, can get them ordered. In case of tubes, it is friendlier to use flat-bottomed ones. However, these kinds of tubes are not readily available in India.

4
Plant Tissue Culture Media

Plant tissue culture medium is an artificial nutrient supplement of organic and inorganic nutrients used for the cultivation of plant tissues. An appreciation and knowledge of the nutritional requirements and metabolic needs of the cultured plant cells and tissues is invaluable in a decision on the type of media to be used in a specific case. The appropriate composition of the medium largely determines the success of the cultures.

There is a broad spectrum of nutritional requirements for optimal growth of a tissue *in vitro* and may vary with the species and therefore, it is often necessary to modify the medium to suit a particular tissue. Even tissues from different parts may have different requirements for satisfactory growth. As such, no single medium can be suggested as being entirely satisfactory for all types of plant cultures. Hence, a significant factor is the choice of nutritional components and growth regulators. During the past 2 to 3 decades, the need to culture diverse tissues and organs has led to the development of several media.

A defined nutrient media consists of inorganic salts, a carbon source, vitamins and growth regulators. Other components may be added for specific purposes. The Murashige and Skoog's (MS) medium (Murashige and Skoog, 1962) is a classic basal medium used worldwide in the laboratories of universities as well as commercial establishments. This media was basically designed to culture tobacco tissues but it is found appropriate for most of the plant species. It is also recognized worldwide that half the strength of MS media is adequate for culturing most plant species. Apart from MS, other media such as Linsmaier and Skoog's (LS) media (Linsmaier and Skoog, 1965) are the most widely used media especially in procedures where plant regeneration is the objective. The B5 (Gamborg *et al.*, 1968), N6 (Chu, 1978), Nitsch and Nitsch (Nitsch and Nitsch, 1969) media and their derivatives have also been used widely for many plant species with different objectives. Generally, a simple medium contains only inorganic salts and utilizable sugar and for most purposes, it is essential to supplement the medium with vitamins, amino acids and growth regulators in different qualitative and quantitative combinations.

4.1. SOLID AND LIQUID MEDIA

Plant material can be cultured either in a liquid medium or on a medium which has been solidified with the gelling agent. The method employed will depend on the type of culture and its objective.

Solid media are widely used for explant establishment, for routine culture of callus or plant organs, and for the long-term maintenance of cultures. Agar is the most commonly used solidifying agent because of the following advantages:

- It forms gels that melt at ca. 100°C and solidify at ca. 45°C
- Gels are not digested by plant enzymes
- Agar does not react with media constituents

Only the lower surface of the explants grown on solid media comes in contact with the medium. Thus, as growth proceeds there will be increasing gradients in nutrients, growth factors and waste products of metabolism between the medium and the tissues and between one part of the callus and another. Gaseous diffusion into and out of the cells at the base of the organ or tissue may also be restricted by the surrounding medium. Though agar is the most commonly used gelling agent in plant tissue culture media there are other gelling agents also used like gelrite, phytagel, methocel (methyl cellulose), barley starch, sago, tapioca starch, etc.

Liquid medium is essential for suspension cultures, but are now also preferred for critical experiments on the nutrition, growth and differentiation of callus tissues and also for micropropagations work. Larger organs can often be satisfactorily grown in a shallow layer of non-agitated liquid, but many tissues and organs, either small or large, grow well unsupported in a liquid medium, provided it is shaken or moved in some way. The purposes served by agitation include:

- The provision of increased aeration
- The reduction in the effect of plant polarity
- The uniform distribution of nutrients and
- The dilution of toxic explant exudates

In liquid media, if a support is needed to raise any cultures, it is a normal practice to use a filter paper bridge (Plate 9). Filter paper bridge is made out of filter paper of good quality preferably of Whatman brand. This bridge is made just by folding a strip of paper in such a way that it resembles a stool with a flat surface on top. The side surface will be immersed into the liquid media. The width of the filter paper depends on the availability of space in a tube or a culture vessel. The length depends on the height of the media. Other kinds of supporting material could be ingeniously designed by the scientists himself. The material thus selected for supporting tissue cultures in liquid media should withstand heat sterilization. The first author had used a polypropylene mesh like substance called as Declon Blue (U.K.) (Plate 10) and also a material known as Milcap (U.K.) (Plate 11).

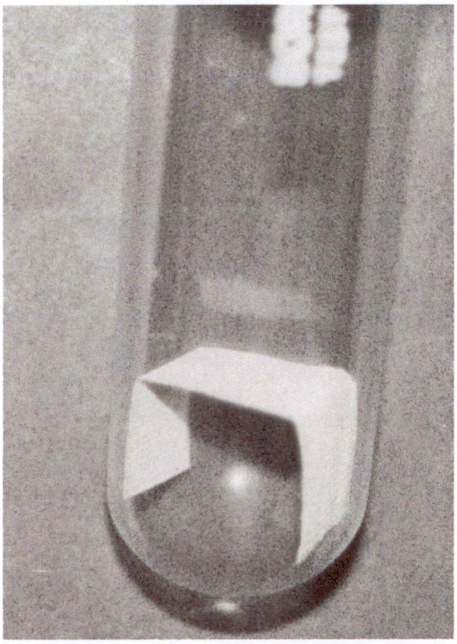

Plate 9: Culture tube with filter paper bridge

Plate 10: Jackfruit plantlet on liquid MS media with Declon Blue as a support

Plate 11: Milcap, a support for micrografts cultured on liquid media

It is observed by the first author during tissue culture of Jackfruit that liquid media supports the growth of Jackfruit tissues splendidly well even in the absence of agitation. For *in vitro* rooting particularly, the results with liquid media were extraordinary in that the root quality was superior. The roots thus obtained were more flexible, in many cases showed increased length, number and growth of secondary roots.

Use of paper bridges to support explants in a liquid media has also been employed to reduce the effect of phenolic exudation from the explants, which otherwise might affect the establishment of the culture. In tissue culture of banana, the second author has experienced that the initiation of cultures from shoot tips were better if the explants were placed on filter paper bridges in liquid media. This probably dilutes the effect of the exuding phenols.

4.2. MEDIA COMPOSITION

The principle components of most plant tissue culture media are:

1. Inorganic nutrients (macro and micro nutrients)
2. Organic supplements
3. Growth regulators
4. Carbon source
5. Gelling agent

4.2.1. Inorganic Nutrients

The mineral nutrients required in concentrations greater than 0.5mmols/L are referred to as macronutrients and those required in concentrations less than 0.05mmols/L are referred to as micronutrients.

4.2.1.1. Macronutrients

The macronutrients include six major elements: nitrogen, phosphorus, potassium, calcium, magnesium and sulphur present as salts that constitute the different media. All of these are essential for plant cell and tissue growth. Culture media should contain at least 25mmols/L nitrates and potassium. However, considerably better results are obtained if the source for nitrogen in the media is contributed by both nitrates and ammonium or any other reduced nitrogen source. The salts normally used to provide the macroelements in tissue culture media also provide sodium and calcium.

***Nitrogen*:** It influences the rate of plant growth and is an essential element in the molecular make-up of nucleic acids, proteins, chlorophyll, amino acids, alkaloids and some plant hormones. Lack of nitrogen develops chlorotic leaves and stunted growth. Excess nitrogen promotes vigorous growth of cultures. Common sources of nitrogen in plant tissue culture media are ammonium (NH_4^+)

and nitrate (NO_3^-) compounds such as through ammonium nitrate and potassium nitrate. It is better to avoid using compounds with chlorine. For example, ammonium chloride or potassium chloride etc., as these would cause a drastic reduction of pH in the media after transfer of cultures. Nitrogen has a great influence on rooting of shoot explants *in vitro*. The first author has done exhaustive studies on the influence of different forms of nitrogen such as ammonium and nitrate and elucidates the correlation these nitrogen forms have with the changes in pH of the media and their effects on induction of rooting *in vitro* (Sathyanarayana, 1993).

***Phosphorus*:** It is found in meristematic and other fast growing tissues. It is a part of DNA and ATP molecules and is an essential element required in photosynthesis and respiration. It also affects plant maturation and root growth. Potassium phosphate (KH_2PO_4) and sodium phosphate ($NaH_2PO_4.H_2O$) are the main sources of phosphorus in plant tissue culture media.

***Potassium*:** Potassium is necessary for normal cell division and promotes meristematic growth. It helps in synthesizing carbohydrates and proteins, chlorophyll, and reducing nitrates. Insufficient potassium results in weak and abnormal plants, sometimes with mottled, curled or dead leaf edges. Potassium nitrate (KNO3) and potassium phosphate (KH_2PO_4) are the common sources of potassium in culture media. KCl is used occasionally.

***Calcium*:** It is an integral component of plant cell walls and helps in the formation of pectin, a substance that bonds cell walls together. It helps control permeability and facilitates the movement of carbohydrates and amino acids throughout the plant. It also promotes root development. Dead shoot or root tips can be a sign of calcium deficiency. Calcium is included in the tissue culture medium as calcium chloride ($CaCl_2. 2H_2O$) or as calcium nitrate ($Ca[NO_3]_2. 4H_2O$).

***Magnesium*:** It is the central element in chlorophyll molecule. It is also important as an enzyme activator. Magnesium deficiency causes older leaves to become chlorotic. Most tissue culture media contain magnesium as magnesium sulphate ($MgSO_4.7H_2O$).

***Sulphur*:** Sulphur is a component of some proteins. It promotes root development and deep green foliage. It is supplied in the media as sulphate compounds.

4.2.1.2. Micronutrients

The inorganic elements required in small quantities but essential for plant cell and tissue growth constitute the microelements. They include iron, manganese, zinc, boron, copper and molybdenum. Of these, iron seems to be more critical and its chelating forms are commonly used in preparing culture media. Iron tartarate and citrate are difficult to dissolve and are frequently precipitated in the medium. This problem may be overcome by using iron chelated to EDTA. The chelating agent EDTA is also known to be one of the causes of formation of genetically aberrant plants. It may be a good idea to use EDTA at lower concentration than the recommended in such plants where genetic aberrations are commonly occurring.

***Iron*:** It is involved in chlorophyll synthesis. It also participates in energy conversions in photosynthesis and respiration as it is reduced from the ferrous to the ferric state. Plants deficient in iron have chlorotic leaves especially between veins giving a striped appearance. In tissue culture media, ferrous sulphate ($FeSO_4.7H_2O$) is often mixed with the sodium salt of ethylenediaminetetraacetic acid (Na_2EDTA) to sequester the iron, thereby making it more readily available to the plants.

***Boron*:** It is presumed to play a role in the movement of sugar, water and hormones. It is also involved in nitrogen metabolism, fruiting and cell division. Lack of boron often results in deterioration of internal tissues, tip die-back, or thick, curled, brittle, chlorotic leaves. Boron is added to the culture media as boric acid (H_3BO_3).

***Chlorine*:** It helps stimulate photosynthesis. Deficiency symptoms are wilted leaves that become yellow and die. Calcium chloride in the media will provide chlorine, often in larger amount than required.

***Cobalt*:** It is a component of Vitamin B12 molecule and is essential for nitrogen fixation. Cobalt chloride ($CoCl_2.6H_2O$) is the source of cobalt in the media.

***Copper*:** Copper is believed to be necessary in energy conversions. It is also involved in chlorophyll synthesis and is a component of enzymes. Copper deficiency results in stunted growth, malformations like twisted and blotched leaves, or die-back of young twigs. In the medium, it is added as cupric sulphate ($CuSO_4.5H_2O$).

***Iodine*:** It is a component of some amino acids. It is often added to the media as potassium iodide (KI).

***Manganese*:** It is an essential element in the chlorophyll membrane. Manganese sulphate provides the necessary manganese for tissue culture media.

***Molybdenum*:** It is believed to convert nitrogen to ammonia and aids in nitrogen fixation. It is required for normal growth and protein synthesis. Its absence is suspected when leaves turn yellow between the veins. It is added to tissue culture media as sodium molybdate ($Na_2MoO_4.2H_2O$).

***Zinc*:** It is an enzyme activator involved in chlorophyll formation as well as production of the auxin indole-3-acetic acid (IAA). Without zinc, roots may be abnormal and leaves may turn mottled bronze or yellow and misshapen. Traces of zinc are included in the media as zinc sulphate ($ZnSO_4.7H_2O$).

4.2.2. Organic Supplements

Vitamins

Plants synthesize vitamins endogenously and these are used as catalysts in various metabolic processes. However, it is necessary to supplement the medium with the required amount of vitamins to achieve the best growth of the tissue. Vitamin B complex contains essential compounds for plant metabolism and growth. The growth substances found in yeast extract, which was commonly used in culture media in the past, are now identified as Thiamine (Vitamin B_1), Nicotinic acid (Niacin or Vitamin B_3), and Pyridoxine (Vitamin B_6). Pantothenic acid (Vitamin B5) and myoinositol are also used often. Thiamine is the best vitamin required by all the cells and tissues. Other vitamins such as biotin, folic acid, ascorbic acid, riboflavin, vitamin E and p-amino benzoic acid are also used.

Amino Acids

Cultured tissue is normally capable of synthesizing the amino acids required for various metabolic processes. The addition of amino acids to the media however, is important for stimulating cell growth and establishing cell lines. The commonly used amino acids are L-glutamine, L-asparagine, L-glycine, L-argine and L-cystein.

Other Organic Supplements

Culture media are also often supplemented with a variety of organic extracts, which have constituents of an undefined nature. These include protein hydrolysates, coconut milk, yeast, malt extract, ground banana, orange juice, tomato juice, etc. However, the use of these natural extracts should be avoided as far as possible as it may affect the reproducibility of the results.

4.2.3. Growth Regulators

In addition to the nutrients, it is generally necessary to add one or more of the growth substances such as auxins, cytokinins, gibberellins and abscisic acid to support good growth of the tissues and organs. Generally, the growth, differentiation and organogenesis of tissue become feasible only on the addition of one or more of these classes of these hormones to the medium. The advent of auxins and cytokinins way back in 1930s had paved the way for a revolution in nursery production and management of plants. Similarly, the spectacular success in tissue cultures should be owed to the findings of these and their use in culture. In fact, the type of growth regulator used in the media determines the kind of response obtained in tissue cultures. Hence, tissue culture is unimaginable in the manner we know without the growth regulators. However, the requirement for these substances varies considerably with the tissue and their endogenous levels.

Auxins

The concept of chemical messengers in plants was proposed by Charles Darwin and his son, Francis, in 1881, who spent time looking at the phenomenon of phototropism in wheat seedlings. Plant shoots are positively phototropic. When a seedling is illuminated from the side, the shoot will bend towards the light. This directional growth makes sense, since plants need light for photosynthesis. However, the Darwins found that if the coleoptile of the wheat seedling was removed, the plant no longer curved towards light. They performed a number of experiments and determined that a chemical located in the coleoptile travelled to the region of elongation and affected a differential elongation of cells furthest from the light sources. The chemical was subsequently studied and named **auxin** by Frits Went in 1926.

Auxins are a general class of the plant hormones, which can cause cell elongation. Chemically, auxin is indoleacetic acid (IAA), which is synthesized from tryptophan. There are a number of synthetic "auxins" too. A primary site of auxin production is the apical shoot meristem but generally help in root initiation and growth. Auxin moves down the stem parenchyma cells by polar transport (auxin becomes negatively charged) using proton pumps, an ATP requiring process.

The auxins mainly used for media supplementation are Indole-3-butyric acid (IBA), Naphthalene acetic acid (NAA) and 2, 4-Dichlorophenoxyacetic acid (2, 4-D). Of these, IBA and NAA are widely used for rooting and in interaction with a cytokinin, for shoot proliferation. 2,4-D and 2,4,5-T are very effective for the induction and growth of the callus. There are others as well, which have proved to be effective particularly in plant cell culture. These include p-chlorophenoxyacetic acid (PCPA), 2-methyl, 4-chloro phenoxyaceticacid (MCPA), etc.

In tissue cultures, auxins have been commonly used for cell division and root differentiation. In nature, the hormones of this group are mainly involved with activities such as elongation of stem and internodes, tropism, apical dominance, abscission, and rooting.

Cytokinins

Cytokinins are a group of phenyl urea derivatives of adenine, one of the molecules in DNA. The first cytokinin was chemically isolated in 1913. Cytokinins have been studied using coconut endosperm for a number of years since 1940s. However, cytokinins were not chemically identified until 1963 in corn. The cytokinin identified from corn was called zeatin.

Cytokinins are mainly concerned with cell division, modification of apical dominance and shoot differentiation in tissue culture. They have their mode of action on the shoot but are produced in the root. The most frequently used cytokinins are benzyl amino purine (BAP) or 6-benzyl adenine (BA), 6-β-β-dimethylaminopurine (2-ip), N-(2-furfurylamino)-1-purine-6-amine (Kinetin) and 6-(4-hydroxy-3-methyl-trans-2 butyl amino purine) (Zeatin). Zeatin and 2-ip are naturally occurring cytokinins while BA and Kinetin (Kn) are synthetically derived cytokinins. They are generally dissolved in dilute HCl or NaOH.

The ratio of auxins and cytokinins is important with respect to morphogenesis in the culture system. For embryogenesis, callus initiation, and root initiation, the requisite ratio of auxin to cytokinin is high while the reverse leads to shoot proliferation.

Gibberellins

The discovery of gibberellins is credited to Ewiti Kurosawa who found that a fungus was responsible for abnormal growth of rice seedling, called the "foolish seedling" disease. The fungus secreted a chemical that caused the rice plants to grew abnormally long, and then collapse from weakness. The fungus was *Gibberella fujikuroi*, hence the hormone named as Gibberellin.

Gibberellins are produced in roots and younger leaves. Of the more than 100 known gibberellins, GA_3 is the most commonly used one. It promotes the growth of cell cultures and induces dwarf and stunted plants to elongate. For meristem culture, after shoot primordia formation, gibberellins are used in plant regeneration and elongation. It is also used to stimulate normal development of the plantlets from *in vitro* formed adventitious embryos. GA_3 is readily soluble in cold water upto 1000mg/L.

Abscisic Acid

Abscisic acid (ABA) is a hormone that functions by inhibiting growth activities in times of environmental stress rather than by promoting growth. It often serves as an antagonist to the other growth promoting hormones in plants. Abscisic acid, which is also synthesized from mevalonic acid, derived its name from the erroneous belief that it promoted the formation of abscission layers in leaves and fruits.

ABA functions as an endogenous regulator of abscission in plants. It has antiauxin effects. Abscisic acid has been found to be useful in embryo culture and somatic embryogenesis in low concentrations. Abscisic acid is heat stable but light sensitive.

4.2.4. Source of Carbon and Energy

Plant cells and tissues in the culture medium lack autotrophic ability and therefore need external carbon for energy. The addition of external carbon source to the medium enhances proliferation of cells and regeneration of green shoots.

4.2.5. Solidifying Agents

Gelling or solidifying agents are commonly used for preparing semi-solid or solid tissue culture media. In static liquid culture, tissues or cells become submerged and die due to the lack of oxygen. To avoid this, the tissue culture media are gelled with agar, a polysaccharide, obtained from a seaweed which supports the tissues grown in static condition.

***Agar*:** Agar is the most commonly used solidifying agent. Marine red algae contain the structural polysaccharide **agar**, which consists of 2 components, **agarose** and **agaropectin**. Agarose is composed of alternating D-galactose and 3,6-anhydro-L-galactose with side chains of 6-methyl-D-galactose residues. Agaropectin is like agarose but additionally contains sulphate ester side chains and D-glucuronic acid. The tertiary structure of agarose is a double helix with a so-called threefold screw axis. The central cavity of this double helix can accommodate H_2O molecules. Agarose and agaropectin readily form gels that contain high amounts of H_2O (up to 99.5%). When agar is mixed with liquid, it forms a gel that melts at about 100°C and solidifies at about 45°C. Other benefits of agar are that it does not react with any components of the medium and it is not digested by enzymes from the plant tissue. Agar contains impurities, such as inorganic salts, organic compounds, phenolics, and long chain fatty acids; amounts and types vary depending on the manufacturer. These compounds usually do not interfere with culture response. If necessary, agar can be washed to remove inhibitory impurities. Agar does not gel well under acidic conditions (pH <4.5). The inclusion of activated charcoal in media may also inhibit gelling of agar.

The agar concentrations commonly used in plant tissue culture media range between 0.5% and 1%; these concentrations yield a firm gel at typical media Hs. The concentration of agar may be critical to plant response in culture. Medium that is too soft may produce hyperhydric (abnormal, glassy-looking) tissue while medium that is too hard may cause reduced growth.

Solidified medium is widely used because such cultures are convenient to maintain and is quite satisfactory for most purposes. However, agar is not an essential component of the nutrient medium. Single cells and cell aggregates can be grown as suspension cultures in liquid medium which should however, be regularly aerated either by bubbling sterile air or by gentle agitation. The protoplasts culture for which a thin film or small drops of liquid medium are used, however, may be maintained in a static state.

***Agarose*:** When greater purity is needed, agarose may be used. Agarose is extracted from agar leaving behind agaropectin and its sulphate groups. Because of the additional purification, agarose is considerably more expensive than agar. Agarose also has higher gel strength than agar and thus less is required for solidification of media. Agarose is used in situations where the impurities of agar are a major disadvantage such as in protoplast culture.

***Gelrite*:** Gelrite™ consists of a polysaccharide produced by the bacterium *Pseudomonas elodea*. Medium solidified with Gelrite is always clear, but agar-solidified medium does not have this advantage. Consequently, contamination is more easily detected at an early stage. Impurities in Gelrite contain inorganic ions, but no organic compounds. Gelrite requires more stirring than agar when being added to media. Unlike agar, Gelrite cannot be reheated and gelled successfully. One limitation of Gelrite is that the concentration of divalent cations such as calcium and magnesium ions must be within the range of 4-8 mM/L. Concentrations of these two ions beyond (either less than or greater than) this range do not allow the gelling of media. Gelrite™ may also produce hyperhydric plants when used at low concentrations.

***Phytagel*:** Phytagel™ is an agar substitute produced from a bacterial substrate composed of glucuronic acid, rhamnose and glucose. It produces a clear, colourless and high-strength gel, which aids in detection of microbial contamination (Plate 12). Phytagel provides an economical alternative to agar as a gelling agent. It is used at a concentration of 1.5-2.5 g/L. To prevent clumping, Phytagel should be added to rapidly stirring culture medium, at room temperature. Hyperhydricity may also be a problem with this gelling agent.

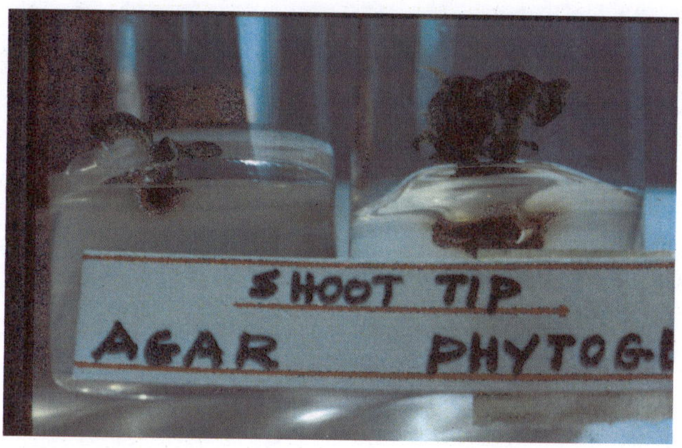

Plate 12: Agar and phytagel used as solidifying agent

***Alternate gelling agents*:** Though agar is the universally appreciated gelling agent used in plant tissue culture, other alternative gelling agents are also used like starch from different sources. To make tissue culture more competitive as far as commercialization is concerned, the need for newer and cheaper gelling agents have become imperative. Gelling agents like tapioca (extracted from tubers of cassava) and sago (a starchy material obtained from the stem pith of

Metroxylon sagu) have shown their successful use in the micropropagation of potato (Kedisso and Sathyanarayana, 1998). These gelling agents could be used at a concentration of 10-15% (Plate 13).

The selection of a gelling agent for specific plants is generally empirical. For unknown reasons, tissues of some species grow more vigorously on one gelling agent than on another. Another major consideration is the degree of hyperhydricity induced in a species by the different gelling agents. One potential way to overcome this is to combine agar and either Gelrite or Phytagel in the medium.

Plate 13: *In virto* microtubers produced in media gelled with tapioca and agar

4.2.6. Miscellaneous Constituents

Antibiotics

Addition of antibiotics to culture media is generally avoided because their presence in the medium retards the tissue growth. However, some plant cells have a systemic infection of microorganisms. To prevent the growth of these microorganisms, it is essential to enrich the media with the antibiotics.

Activated Charcoal

Activated charcoal or activated carbon is an amorphous form of carbon. The chemical nature of amorphous carbon, combined with a high surface area and porosity, makes it an ideal medium for the adsorption of many organic and inorganic molecules from a culture medium and this has been used in a variety of tissue culture systems. It may remove toxic contaminants from agar and secondary products secreted from the cultured tissues, thereby preventing the browning of explants. It may be especially useful in rooting medium. However, activated charcoal adsorbs not only toxic compounds, but also growth regulators and other compounds that are added to the medium. Activated charcoal is usually acid-washed prior to addition to the culture medium at a concentration of 0.5-3.0%.

The addition of activated charcoal to culture media is reported to promote growth and differentiation in orchids, carrot and tomato tissue cultures.

Antioxidants

An antioxidant is an electron donor, which inhibits the oxidation of labile substrates with high stochiometric efficiency. The tendency of dissolved compounds to be oxidized or reduced depends on the oxidation-reduction potential (redox) of the solution. Antioxidants mainly act as reducing agents, which lower the redox potential of solutions and thus are effective in preventing the blackening of isolated plant tissues or plant extracts. Ascorbic acid and citric acid are commonly used in plant tissue culture. A solution of ascorbic acid and citric acid is commonly used for rinsing freshly isolated explants to delay the browning process.

Antioxidants have also been incorporated into tissue culture media to reduce excessive browning of media. However, ascorbic acid is heat labile and hence, should be filter sterilized before addition and not autoclaved. Citric acid can also be added to the media, but this is frequently for purposes other than the prevention of blackening because it can act both as a buffer and as a nutrient.

Other antioxidants which have been tried in tissue culture are glutathione, L-cysteine, dithiothreitol etc.

pH of the Medium

The relative acidity or alkalinity of a solution is assessed by its pH. This is a measure of the hydrogen ion activity. The pH of the medium must be such that it will not disrupt the function of plant cell membranes or the buffered pH of the cytoplasm. The pH of the medium governs whether the salts will remain in a soluble form, influences the uptake of medium ingredients and affects the gelling efficiency of agar. Hence, the testing of the pH is an integral activity during the preparation of tissue culture media. Failing to read and adjust the pH of the media before culturing could be a ghastly mistake as this may affect the growth of cultures.

The pH of the medium is normally adjusted in the range of 5.6 to 5.8 but different plants may require different pH for optimum growth. The pH is most commonly adjusted using NaOH (sometimes KOH) or HCl when all medium components have been mixed just prior to autoclaving.

Table 2: Media for Plant Tissue and Cell Cultures (mg/L)

Components	*Murashige-Skoog (1962)*	*White (1963)*	*B5-Gamborg (1968)*	*Nitsch (1951)*	*Heller (1953)*	*Schenk-Hildebrandt (1972)*	*Nitsch-Nitsch (1967) (1982)*	*Woody Plant Media*	*Kohlenbach-Schmidt (1975)*	*Knop (1865)*
$(NH_4)_2SO_4$	—	—	134	—	—	—	—		—	—
$MgSO_4 \cdot 7H_2O$	370	720	500	250	250	400	125	370	185	250
K_2SO_4								990		
Na_2SO_4	—	200	—	—	—	—	—	—	—	
KCl	—	65	—	1,500	750	—	—		—	—
$CaCl_2 \cdot 2H_2O$	440	—	150	25	75	200	—	96	166	—
$NaNO_3$	—	—	—	—	600	—	—		—	—
KNO_3	1,900	80	3,000	2,000	—	2,500	125		950	250
$Ca(NO_3)_2 \cdot 4H_2O$	—	300	—	—	—	—	500	556	—	1,000
NH_4NO_3	1,650	—	—	—	—	—	—	400	720	—
$NaH_2PO_4H_2O$	—	16.5	150	250	125	—	—		—	—
NH_4HPO_4	—	—	—	—	—	300	—		—	—
KH_2PO_4	170	—	—	—	—	—	125	170	68	250
$FeSO_4 \cdot 7H_2O$	27.8	—	27.8	—	—	15	27.85	27.8	27.85	—
Na_2EDTA	37.3	—	37.3	—	—	20	37.25	33.6	37.25	—
$MnSO_4 \cdot 4H_2O$	22.3	7	10 (1 H_2O)	3	0.1	10	25		25	—
$MnSO_4 \cdot H_2O$								22.3		
$ZnSO_4 \cdot 7H_2O$	8.6	3	2	0.5	1	0.1	10	8.6	10	—
$CuSO_4 \cdot 5H_2O$	0.025	—	0.025	0.025	0.03	0.2	0.025	0.25	0.025	—
H_2SO_4	—	—	—	0.5	—	—	—		—	—

Contd...

Contd...

Components	*Murashige-Skoog (1962)*	*White (1963)*	*B5-Gamborg (1968)*	*Nitsch (1951)*	*Heller (1953)*	*Schenk-Hildebrandt (1972)*	*Nitsch-Nitsch (1967) (1982)*	*Woody Plant Media*	*Kohlenbach-Schmidt (1975)*	*Knop (1865)*
$Fe_2(SO4)_3$	—	2.5	—	—	—	—	—		—	—
$NiCl_2.6H_2O$	—	—	—	—	0.03	—	—		—	—
$CoCl_2.6H_2O$	0.025	—	0.025	—	—	0.1	0.025		—	—
$AlCl_3$	—	—	—	—	0.03	—	—		—	—
$FeCl_3.6H_2O$	—	—	—	—	1	—	—		—	—
$FeC_6O_5H_7.5H_2O$	—	—	—	10	—	—	—		—	—
KI	0.83	0.75	0.75	0.5	0.01	1.0	—		—	—
H_3BO_3	6.2	1.5	3	0.5	1	5	10	6.2	10	—
$Na_2M_0O_4.2H_2O$	0.25	—	0.25	0.25	—	0.1	0.25	0.25	0.25	—
Sucrose/Glucose	30,000	20,000	20,000	50,000	20,000	30,000	20,000-	20,000	10,000	—
	—	—	—	or	—	—	30,000		—	—
				36,000			000			
Myo-Inositol	100	—	100	—	—	1,000	100	100	100	—
Nicotinic Acid	0.5	0.5	1.0	—	—	0.5	5	0.5	5	—
Pyridoxine HCl	0.5	0.1	1.0	—	—	0.5	0.5	0.5	0.5	—
Thiamine HCl	0.1-1	0.1	10	1	1	5	0.5	1.0	0.5	—
Ca-Pantothenate	—	1	—	—	—	—	—		—	—
Biotin	—	—	—	—	—	—	0.05		0.05	—
Glycine	2	3	—	—	—	—	2	2.0	2	—
Cysteine HCl	—	1	—	10	—	—	—		—	—
Folic Acid	—	—	—	—	—	0.5		0.5	—	
Glutamine	—	—	—	—	—	—		14.7	—	

4.2.7. Types of Media

Wide ranges of culture media differing in their elemental compositions have been standardized. Some of the most widely used culture media and their composition are given in Table 2.

MS Medium

This is the most popular and broad-spectrum medium designed by Murashige and Skoog (1962). This revised medium contains 40mmols nitrogen as nitrates and 20mmols nitrogen as ammonium. The total ion concentration of nitrogen is 60.13mmols, which was five fold more than that of the earlier media used and 19 times higher than White's medium. Potassium was also higher with a concentration of 20mmols. The composition of MS media is given in Table 3.

B5 Medium or Gamborg's Medium

This medium differs from other media in having much higher levels of nitrate and potassium, but has low concentration of ammonia.

White's Medium

White's medium is one of the earliest plant tissue culture medium and includes all the necessary nutrients, but with less concentration of nitrites. The prominence was given to magnesium sulphate. In the original White's medium (1943), iron was added as $Fe_2(SO_4)_3$, but Street *et al.* replaced it with $FeCl_2$. White's medium is widely used for root culture.

Nitsch's Medium

This medium is especially suitable for the culture of anther so as to produce haploid plants. It differs from MS medium in having half of its macronutrients and the concentration of microelements and others remain same as that of MS medium.

Table 3: Composition of Murashige and Skoog Medium (1962)

Components	*Concentration (mg/L)*
Macronutrients	
NH_4NO_3	1,650
KNO_3	1,900
$CaCl_2 . 2H_2O$	440
$MgSO_4 . 7H_2O$	370
KH_2PO_4	170
Fe NaEDTA	36.7

Contd...

Contd...

Micronutrients	
H_3BO_3	6.2
$MnSO_4 . 4H_2O$	22.3
$ZnSO_4 . 7H_2O$	8.6
KI	0.83
$Na_2MoO_4 . 2H_2O$	0.25
$CuSO_4 . 5H_2O$	0.025
$CoCl_2 . 6H_2O$	0.025
Vitamins	
Thiamine HCl	0.1
Myo-inositol	100.0
Nicotinic Acid	0.5
Pyridoxine HCl	0.5
Glycine	2.0
Carbon Source	
Sucrose/Glucose	30,000

Woody Plant Medium

Woody Plant Medium, popularly referred to as WPM was formulated by McCown and Lloyd (1980). The main feature of Woody Plant Medium is that for a post-Murashige and Skoog formulation, its total ionic concentration is low. This is consistent with other media devised for woody species, but sulphate is present at a higher than average concentration. Woody Plant Medium is increasingly used for the propagation of ornamental shrubs in commercial laboratories. WPM could be a basal media for *in vitro* rooting purposes owing to its low salt concentration (Plate 14).

Lindemann Orchid Basal Salts Mixture

It is one of the well known media for orchid culture. It has a lower salt strength. This media contains 15.2 mM nitrogen in the form of ammonium, whereas only 4.2 mM nitrogen in the form of nitrate. Wherever lower concentrations of salts are required for a particular plant species culture or higher ratio of ammonium to nitrate is required this media could be used. Generally, this media may help in better rooting of micro shoots (Plate 14).

Plate 14: Rooting of jackfruit microcuttings in different commer-cially available media

4.2.8. Preparation of Plant Tissue Culture Media

It is a standard laboratory procedure to make stock solutions of all the chemicals needed for the preparation of media. As you will read later in this chapter, media could be prepared by mixing the stock solutions in the required amounts as and when necessary. Generally during media preparation, the required amount of sugar is first dissolved in double distilled water, stock solutions are added and the media is gelled by boiling after the addition of a gelling agent. Thereafter, the media is dispensed into culture vessels and autoclaved. But this method results in the heating of media components twice, during dissolution of the gelling agent and also during autoclaving, which might lead to the disintegration of media constituents. An alternative method is to mix the stock solutions in half the original volume of the media and dissolve the gelling agent in the remaining amount of double distilled water. These two solutions are then mixed together while dispensing into the culture bottles and then autoclaved.

(*a*) Preparation of Stock Solutions

Stock solutions of major salts, minor salts and vitamins are usually prepared separately and stored in the refrigerator. For preparing the stock solutions and media, distilled or demineralized water and chemicals of high purity should be used.

For the preparation of stock solutions, the chemicals are dissolved in distilled or demineralized water. Adding one compound at a time will usually avoid precipitation. Dissolving inorganic nitrogen sources of the major salts first will avoid precipitation between phosphate and calcium sources when added subsequently. Dissolving the calcium salts separately before adding will also help to avoid precipitation.

The kinds and quantities of stock solutions being prepared vary with the kind of media, size of the operation, and preferences. Usually, the stock solutions are prepared in 10X and 100X concentrations. The stocks can consist of groups of chemicals or nearly complete media. All the stock solutions, after preparation, are stored in proper plastic or glass bottles under refrigeration. The stock solutions are preferably stored in compartments of the refrigerator where the temperature ranges from 6 to 10°C. The iron stock, which is usually prepared separately, should be stored in amber coloured bottles. Separate stock solutions are prepared for each growth regulator as well by dissolving it in a very minute quantity of the appropriate solvent (Table 4) and making up the final volume with distilled water.

Note: Making a stock solution of plant hormones may not be as easy as one would like to. It is the experience with students often landing in problems in dissolving different growth regulators. Every growth regulator has a specific solvent to be used. The problem of dissolving in water to make a growth regulator solution may also be due to the quality of water used, the quality of solvent and also more particularly, the quantity of growth regulator to be dissolved. Dissolving a higher amount of growth regulator say, dissolving 100mg or more to prepare solutions of over 100ppm concentration is a difficult task. Though initially, the growth regulator powder dissolves in little amount of alcohol, later, precipitates with the addition of water for making up the solution. Hence, a higher concentration of growth regulator especially auxins will only be achieved with 50 or 100% alcoholic solution.

Table 4: Solvents for Growth Regulators

Compound	*Abbreviation*	*Solvent*
Cytokinins		
Benzyladenine/Benzylaminopurine	BA/BAP	0.5N HCl or 95% ethanol
Isopentenyladenine	2-Ip	0.5N HCl or 95% ethanol
Kinetin	KIN	0.5N HCl or 95% ethanol
Zeatin	ZEA	0.5N HCl or 95% ethanol
Auxins		
Indole-3-acetic acid	IAA	95% ethanol or 1M KOH/NaOH
Indole-3-butyric acid	IBA	95% ethanol or 1M KOH/NaOH
1-Naphthalene acetic acid	NAA	95% ethanol or 1M KOH/NaOH
2, 4-Dichlorophenoxy acetic acid	2, 4-D	95% ethanol or 1M KOH/NaOH
2, 4, 5-Trichlorophenoxy acetic acid	2, 4, 5-T	95% ethanol or 1M KOH/NaOH
Pichloram	PIC	95% ethanol or 1M KOH/ NaOH
Others		
Thidiazuron	TDZ	95% ethanol
Silver nitrate	$AgNO_3$	95% ethanol

As a rule, before using the stocks, the bottles must be shaken gently and if any of the solutions show a suspension or a precipitate or a biological contaminant, they should be immediately discarded.

The components of stock solutions for MS media are given below:

Stock solution 1 (MS-A): Macronutrients (10% concentration sufficient to make 1 L of MS media).

The following chemicals are dissolved in 1 litre of distilled water.

NH_4NO_3	16.5 gm
KNO_3	19.0 gm
$CaCl_2.2H_2O$	4.4 gm
$MgSO_4.7H_2O$	3.7 gm
KH_2PO_4	1.7 gm

Stock solution 2 (MS-B): Micronutrients (100% concentration sufficient for 1 L).

The following chemicals are added to 1 litre of distilled water.

$MnSO_4.4H_2O$	2230 mg
$ZnSO_4.4H_2O$	860 mg
H_3BO_3	620 mg
KI	83 mg
$Na_2MoOH.2H_2O$	2.5 mg
$CuSO_4.5H_2O$	2.5 mg
$CaCl_2.6H_2O$	2.5 mg

Stock solution 3 (MS-C): Fe-EDTA (100% concentration sufficient for 100 ml)

Add the following to 100 ml of distilled water.

$FeSO_4.7H_2O$	278 mg
Na_2 EDTA.$2H_2O$	373 mg

First dissolve EDTA and then iron. Store in a dark bottle at 0-4°C.

Stock solution 4 (MS-D): Vitamins (100% concentration sufficient for 100 ml).

Add the following vitamins to 100 ml of distilled water.

Glycine	20 mg
Nicotinic acid	5 mg
Pyridoxin HCl	5 mg
Thiamine HCl	1 mg
Myo-inositol	1000 mg

(*b*) Preparation of Media

A standard protocol for media preparation is described below. This generalized approach can be adapted to prepare any medium.

- Required quantity of sucrose is weighed and dissolved in distilled water (about three-fourths of the final volume) with the aid of a magnetic stirrer.
- Appropriate quantities of the various stock solutions, including growth regulators and other special supplements are added to it.
- After all the ingredients have been added and dissolved, with the exception of the gelling agent, add distilled water to height just below the final volume.
- Check and adjust the pH to the required value by adding dropwise, while stirring, either 0.5M NaOH or 0.5NHCl as needed.
- After adjusting the pH, bring the solution to the final volume by using a graduated cylinder.
- For gelled media, weigh out the needed amount of agar or gelling agent and add to the slightly warm media. Bring the media to boil on a heater or water bath for proper dissolution of the agar. This step is not necessary for a liquid medium.
- Pour the media into desired culture vessels. About 10 ml of the medium can be dispensed into a 150 × 25mm culture tube or 25-30 ml into glass bottles.
- Autoclave the media at 121°C at 105kPa for 15-20 min. The medium is allowed to cool at room temperature and is then stored at room temperature.

Media could also be prepared by another method. In this method, media is prepared in two stages, unlike the earlier method, where all the ingradients including agar were added together. In this method, Nutrient solution and Agar solution are prepared separately and later mixed together as explained further (This method was being followed in some laboratories including that of Unit of Advanced Propagation Systems, Wye College, University of London.)

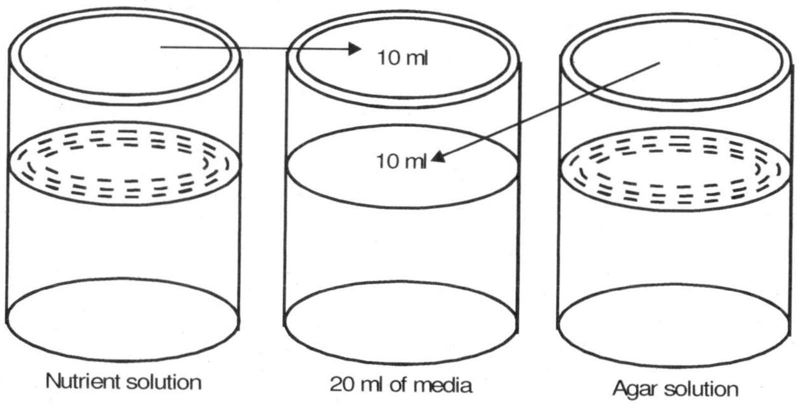

Nutrient solution was first prepared by dissolving the required amounts of powdered formulation of MS media (or any medium), sucrose and growth regulators in less than half the final volume (adjust its pH). Agar solution was prepared by dissolving (by bringing to the boil) the total required amount of agar in the remaining half of the final volume of water. The nutrient solution was first dispered into jars or tubes followed by equal amounts of agar solution. Though this method is comparitively little cumbersome, the nutrient solution is subjected to boiling only once, that is, during autoclaving.

Note:

1. Always establish a step-by-step routine for preparing a medium. This routine will ensure that the medium is identical in composition each time.
2. **Add each component according to the list of ingredients and check it off on the list as it is added. The checklist prepared should contain all the procedures to be carried out for the preparation of the media including the pH. Any lapse in preparing media would cost dearly, wasting the valuable time spent on the preparation of media. If you have cultured without knowing the mistake made, the results obtained could be negative or if it is an experiment, mind you, you are interpreting the results wrongly. By mistake, if you have forgotten say, to add sucrose or growth regulator, all the efforts are absolute waste of time and effort.**
3. Measure each component carefully by using pipette or graduated cylinder of appropriate size. For example, 1.0 ml solution should not be measured in a 10 ml graduated cylinder, but use of 1.0 ml or 5.0 ml pipette will be more appropriate.
4. The medium can be distributed in culture vessels before autoclaving or the containers and the medium can be autoclaved separately. If pre-sterilized, unautoclavable plastic culture vessels are to be used, the medium may be autoclaved in flasks. The medium is then allowed to cool to around 60°C before pouring it into the vials under aseptic conditions.
5. Heat labile compounds such as L-glutamine and gibberellic acids should be filter sterilized and added to the autoclaved media. Filter sterilization is carried out by first dissolving the compound in less than the final volume. After filtration into a sterile container, sterile distilled water is added to bring the solution up to volume under sterile conditions. The filter sterilized compounds are added to the autoclaved media when cooling but not yet solidified, warm to touch but not hot. Mix thoroughly by swirling. (Make sure that the volume of the autoclaved medium plus the volume of the filter sterilized compounds sums to the final volume of the prepare medium).
6. When preparing a solid medium in culture tubes, it is desirable to make slants by keeping the tubes tilted during cooling. Such slants provide a

larger surface area for the tissue growth. It is also easier to photograph cultures grown on such slants.

4.2.9. Common Problems Encountered

Contamination: It can occur in stock solutions of vitamins and growth regulators as well as in prepared media, which may be inadequately sterilized. Any media with actual or suspected contamination should be autoclaved as soon as possible without opening the container.

Precipitation of the salts: Another problem encountered in media preparation is the formation of insoluble precipitates. The main causes of precipitation in the medium are the complexes formed between calcium, phosphate and magnesium. This problem can be avoided if the compounds are added in the order of nitrogen compounds, then the magnesium compounds followed by the calcium compounds and then the phosphate compounds. Each compound should be completely dissolved before the next one is added. The formation of precipitates can also be avoided if the calcium compounds are dissolved separately and added as a solution rather than the salt.

Improper setting of the media: The gel consistency may be too soft at times. This could be due to the following reasons:

- The pressure of the autoclave may have been set too high or the pressure stays high for much longer than expected due to lack of pressure valve control.
- The temperature of the autoclave may have been set too high.
- The duration of autoclaving may be too long.
- The pH of the medium is incorrectly set too low. pH below 4.0 will produce a runny gel.
- Gel concentration is too low.
- The quality of agar used also play a role in setting media. Though normally, good agar should set at 6% strength, the agar available in the market sets anyway between 8-10%. Hence, lower percentage agar normally results in runny gel.
- Type of sugar used. For example, Fructose makes a softer gel.

4.2.10. Preparation of 1 L of MS Medium (full strength) from Stock Solutions

1. To 600 ml of distilled water (with continuous stirring).
   (*a*) Add 100 ml of macronutrient stock solution 1.
   (*b*) Add 10 ml of micronutrient stock solution 2.
   (*c*) Add 10 ml of Fe-EDTA stock solution 3.
   (*d*) Add 10 ml of vitamin stock solution 4.
2. Add cytokinin and /or auxins as necessary.

3. Add sucrose (usually 30 g/ L). Make volume up to 1 L with distilled water.
5. Adjust pH to 5.6 - 5.8.
6. Add agar (usually 6-10 g/ L) and dissolve.
7. Dispense into suitable culture vessels and autoclave.

4.2.11. Preparation of Half Strength MS Media (1L) from Stock Solutions

1. To 600 ml of distilled water with stirring.
   (*a*) Add 50 ml of macronutrient stock solution 1.
   (*b*) Add 5 ml of micronutrient stock solution 2.
   (*c*) Add 5 ml of Fe-EDTA stock solution 3.
   (*d*) Add 5 ml of vitamin stock solution 4.
2. Add cytokinin and /or auxins as necessary.
3. Add sucrose (30 g/L).
4. Make up volume to 1 L with distilled water.
5. Adjust pH to 5.6 - 5.8.
6. Boil the media to add agar (8 g/L).
7. Dispense into baby jar bottle (30 ml/bottle) and autoclave.

Keep a check list- "An episode of humor": It is our experience and I am sure of every one who have prepared media, that one experiences a strange feeling of self doubt, whether or not a particular chemical / stock / ingredient was added. Even more scaring is to nurture a doubt after media was autoclaved: whether pH was adjusted or not. The repercussion of these doubts or genuine mistake of having forgotten to add an ingredient could be agonizing, as it could ruin one's experiment. With serious researchers or students, who are hard pressed for time, a mistake of this nature proves ghastly and extremely depressing especially coming to know the mistake when cultures respond strangely. I (first author) would like to recall a memorable event happend during my PhD at the Laboratory of Unit for Advance Propagation Studies at Wye College (Imperial College, London). It was a busy day, precisely a forenoon, I guess. The Laminar Air flow chambers were fully occupied by the research scholars. The scholars were a heterogeneous group of different nationalities such as India (myself), Bangladesh, Brazil, Kuwait, Columbia, UK, Zimbabwe, etc. Everybody was pretty engrossed in their culture transfer work, when a "jolly go lucky" Brazilian student entered the culture transfer room and occupied his designated (booked) LAF (normally, his entry used to go unnoticed by other students). He, promptly uncovered the brown paper covering a tray of autoclaved culture bottles. While unwrapping, he called the attention of everybody and very jubilantly announced:

"Hey! Listen every body, this is my last experiment". (Everyone turns towards the Brazilian)

"This is the last tray of the media, I would ever culture, believe me! I am sick and tired of this job".

Mr. Rabbani (a Bangladeshi) and Andrew Matibiri (Zimbabwean) sitting before LAF's, situated either side of me, murmured something suggesting on the contrary. No sooner we were back to our work, a huge "cry" of disgust and anger came from Pedro Moura Costa (the Brazilian). What! (Was a collective question of all the fellow mates). Oh! Sh--, I forgot to add agar, man!

"Trust Pedro Moura Costa to do such a mistake" were the collective comments of student scientists at culture transfer room.

Put yourself in Pedro's shoes.

I hope Pedro's takes my narration sportingly.

Hence, it is wiser to make **'a check list'** of do's and properly ticked off after an activity (addition of ingredients, checking pH, etc.) is over. If you nurture a doubt especially when your cultures have responded differently after a period, you can fall back on your check list. The students and researchers, how much ever experienced they are, everytime they set out to prepare media (more particularly when preparing media for different treatments) should consider the day as another day and feel as though they are doing for the first time. Students should be calm and composed and should never allow themselves to be distracted.

5

STERILIZATION TECHNIQUES

Plant tissue culture media, which contain a high concentration of sugars, support the growth of many microorganisms like bacteria and fungi. The initial explant is the major source of infestation, but re-infestation is possible at any stage of the culture process. Generally, the microorganisms are universally present on and within the plant tissues, but the dry conditions and competing organism may keep them under control. On reaching the medium, these microbes generally grow much faster than the cultured tissue and metabolize much of the carbohydrate, producing phytotoxic fermentation products such as ethanol and acetic acid. This will starve the plant tissue of carbohydrates, make certain nutrients unavailable and cause toxic effects on the plant tissues through the secondary products produced. The maintenance of aseptic or sterile conditions are therefore, absolutely essential to maintain a completely aseptic environment and for successful tissue culture procedures.

There are several possible sources of contamination, which includes:

- The culture vessel
- The culture medium
- The explant
- The environment of the transfer room
- The environment of the culture room
- Instruments used to handle the plant material during inoculation and subculturing

Generally, four categories of sterilization are recognized:

1. Sterilization of culture vessels and instruments
2. Sterilization of nutrient media
3. Sterilization of the explants
4. Sterilization of culture room and transfer area

5.1. EQUIPMENTS USED FOR STERILIZATION

1. Autoclave or pressure cooker
2. Laminar flow chambers

3. Hot air oven
4. Filter units

5.1.1. Sterilization of Culture Vessels and Instruments

The culture vessels and other glasswares have to be washed well and dried before sterilization. Traditionally, glasswares were washed with a mixture of potassium dichromate and concentrated sulphuric acid followed by thorough wash under a jet of tap water to completely remove all traces of the acid. Today, detergents especially designed for washing laboratory and plastic wares are available. The glasswares are soaked in the detergent solution for a suitable period and then thoroughly rinsed first under tap water and then with distilled water. To reuse the glasswares that have contaminated tissue or media, it is extremely important to autoclave them without opening the caps so that all the microbial contaminants are destroyed. The washed glasswares and plastic wares are then subjected to sterilization.

(a) Dry heat sterilization

Glass, culture vessels, metal instruments, and aluminium foil can be sterilized by exposure to hot dry air (160-1800°C) for 2-4h in a hot air oven. All items should be properly sealed before sterilization. Sealing with paper is not advisable as they decompose at higher temperature.

Disadvantages

- Poor air circulation
- Slow penetration of heat

(b) Autoclaving

Here, sterilization is done with steam under high pressure. Nearly all the microbes are killed on exposure to the superheated steam in an autoclave. Cotton plugs, guaze, plastic caps or pippetes are autoclaved at 121°C and 15psi for 15-20 minutes. Plastic labwares such as propylene and Teflon can be repeatedly autoclaved.

(c) Flame sterilization

Normally, metal instruments such as forceps, scalpel needles and spatulas are sterilized by dipping in 95% alcohol followed by flaming and cooling. This technique is called flame sterilization. Autoclaving of metallic instruments is generally avoided as they rust and become blunt.

Disadvantages

- During flame sterilization, the major concern should be the safety of alcohol because it is inflammable and if spilled near the flame, will cause an instant fire. To avoid this, steribead sterilizer is used.

5.1.2. Sterilization of Nutrient Media

The nutrient media used in tissue culture are commonly sterilized by autoclaving and by filter sterilization. Distilled water, macro and micronutrients and other stable mixtures are also autoclaved whereas solutions containing thermolabile compounds are filter sterilized.

***Autoclaving*:** Culture media in glass containers sealed with cotton plugs, aluminium foil or plastic caps are autoclaved at 15psi for 15 - 40 minutes from the time the media reaches the required temperature. Exposure time depends on the volume of the liquid to be sterilized (Table 5).

Table 5: Minimum sterilization time for plant tissue culture media (after Biondi, S. and Thorpe, T.A., 1981)

Volume of media for per vessel (ml)	*Minimum sterilization time (min)*
20-50	15
75	20
250-500	25
1000	30
1500	35
2000	40

The sterilization depends upon the temperature and not directly on pressure. The pressure should not exceed 20psi, as higher pressure may lead to the decomposition of carbohydrates and other components of a medium. Care must be taken while cooling the solution. A rapid loss of pressure exceeding the rate of reduction in the temperature will make the liquid boil vigorously. The pressure gauge should be zero before the autoclave is opened.

***Filter sterilization*:** Vitamins, amino acids, plant extracts, hormones and carbohydrates are thermolabile and may decompose during autoclaving. Hence, these are sterilized by filter sterilization. Filter sterilization of complex materials such as coconut milk will require at least two different filters. After an initial coarse filter, use a prefilter of 0.5 µm and a final filter of 0.22 µm. Stock solutions of most heat labile biochemicals can be sterilized using only the final filter of 0.22 µm. The following types of filters are known:

Asbestos and Asbestos-paper disks (Seitz): This consists of an upper cylinder and a lower funnel and the asbestos pad in between. There are different grades of filter pads used. The disadvantage of the pad is that they absorb small amounts of filtrate.

Sintered glass filters: These are made up of finely ground glass, fused together to make the glass granules adhere to one another.

Microfilters: These are used to filter small amounts of fluids. They can be membrane or asbestos. (For example, The syringe filter, The centrifuge filter).

Depth filters: They consist of fibrous or granular materials that have been bonded into a thick layer filled with twisting channels of small diameter. The solution containing microorganisms is sucked through this layer under vacuum, and microbial cells are removed by physical screening or entrapment and by adsorption to the surface of the filter material. Depth filters are made of diatomaceous earth (Berkefield filters), unglazed porcelain (Chamberlain filters), asbestos, or other similar materials.

Cellulose membrane filters: These are less absorptive and can be used to separate viruses. The solution is forced through the filter with a vacuum or with pressure from a syringe or peristaltic pump and collected in previously sterilized containers. Membrane filters remove microorganisms by screening them out much as a sieve separates large particles from small ones. Two types of membrane filters are available:

1. Gradocol membrane, made of cellulose nitrate
2. Cellulose acetate membrane: These are modern and commonly used filters. The membrane is about 120 μm thick and placed in two layers. Bacteria are trapped in the upper layer (0.5 - 1.0 μm) and more perfect than Seitz.

Microwave sterilization of media has also been reported, but may be practical only for smaller quantities of media.

5.1.3. Sterilization of Explants

Surfaces of plant parts carry a wide range of microbial contaminants. To avoid this source of infection, the tissues must be thoroughly surface sterilized before placing it on the nutrient medium. Tissues with systemic fungal or bacterial infection are usually discarded in tissue culture studies.

Sterilization of plant materials is achieved in various ways. The most convenient procedure is to produce the explant materials under aseptic conditions, such as from aseptically germinated seedlings. The tissues to be used are then removed and are then immediately ready for use. The sterilization procedures that are needed for explants produced under non-aseptic conditions vary. It is essential to remove dirt and debris from the plant tissue and should be washed in a detergent solution and rinsed several times in distilled water prior to sterilization. In cases where the explant carries heavy load of microorganisms on its surface, it is better to wash under running tap water for an hour or more.

The sterilization procedures that are needed for plant materials, which are to be directly used as explant sources, can also cause damage to the tissues. Therefore, the concentration of the sterilizing agent and the duration of treatment should be adjusted such as to minimize tissue death. A list of some commonly used sterilents is given in Table 6.

Table 6: Common sterilizing agents used for surface sterilization of explants

Sterilizing Agent	*Concentration (%)*	*Time of Sterilization (min)*
Mercuric chloride	0.01-0.1	0.5-10
Calcium hypochlorite	9–10	5–30
Sodium hypochlorite	2	5–30
Hydrogen peroxide	10–12	5–15
Bromine water	1–2	2–10
Chlorine water	—	—
Silver nitrate	1	5–30
Silver chloride	0.1–1	2–30
Benzalkonium chloride	0.01–0.1	5–20

The outer surfaces of many plants may contain a waxy cuticle layer. It may be necessary to add 2-3 drops per litre (0.01%) of a commercial detergent to the disinfectant to allow better penetration of the waxy layer and of the small pores on the tissue surface. Addition of reagent grade surfactants (Tween-20 (polyoxyethylene sorbitan), Teepol, or Triton X-100) often added in low concentrations to the chemical sterilization solution ensures that the sterilizing agent comes in contact with the entire plant tissue surface.

In general, if the explant is fairly hard and large enough to be easily handled, it can be directly treated with the disinfectant. For instance, in the culture of mature seeds or mature endosperm, the whole seeds or de-coated seeds, respectively, are surface sterilized. However, when immature ovules, embryos or endosperm are to be cultured, the customary method is to surface sterilize the ovary or the ovules as the case may be and dissect out the plant under aseptic condition so that the soft tissues of the explant are protected from the toxic effects of the sterilizing agent. Similarly, for raising culture of delicate shoot apices and pollen grains, shoot buds or flower buds are surface sterilized and the explant is excised aseptically. A careful dissection of shoot apices may give high frequency of healthy cultures even without a surface sterilization treatment.

After surface sterilization treatment, the plant material must be rinsed 3 or 4 times in sterile water to remove all traces of the sterilizing agent. It is our experience that no single formulae or procedure or a protocol would do for all the species. Even for the same species or the same variety, a single formula would not work done at different times. Season, place, the kind of explant used, the time of collection of explant, all will have a decisive role on the successful surface sterilization. In fact, it is the load of microorganisms on an explant, which determines the successful protocol of surface sterilization.

5.1.4. Sterilization of Transfer Area

It is very essential that complete precaution is taken to prevent the entry of any contaminant into the culture vial when its mouth is opened either for subculture or for inoculation of fresh tissue. To achieve this, all transfer operations are carried out under strictly aseptic conditions.

The transfer area is cleaned once or twice a month with a commercial brand of a disinfectant. However, larger transfer rooms are best sterilized by exposure to UV light. The effectiveness of UV radiation is dependent on direct contact it makes with the particles carrying the microorganisms, because they have very less penetrative power. The UV rays act by inducing thiamine dimers in DNA and this interferes with replication of the DNA of organisms. All the sterile transfers are to be done in a laminar flow hood, which is provided with UV sterilizing tubes. More information in this aspect could be gathered in the Chapter 3 on Instrumentation.

5.1.5. Sterile Transfer

The plant materials, which were decontaminated through surface sterilization techniques, as explained earlier, have to be transferred onto the media, which is also made sterile by autoclaving. This transfer of sterile plant material onto sterile media thus has to be necessarily done in a sterile area. Sterile area is normally provided by laminar airflow bench stations. The students or the researcher should ensure that all the activities during transfer have to be perfectly done. Any callousness or negligence on the part of a student or a researcher would show up on the media/culture bottle, which gets contaminated. Right attitude, technique and approach is the right way to accomplish perfect sterile transfer. To proceed with, the operator (student or researcher) should be seated comfortably and assume a posture of sitting, which makes him/her feel comfortable throughout the hours of working. (Gloves, head cover etc). Before the actual transfer of the explants onto the media, wherever there is a facility of UV light is available, UV light should be switched on 15 minutes prior to the plant transfer operation. In the absence of UV light, the airflow could be switched on. In the event of UV light being used prior to the plant transfer, it should be switched off before transfer operation. The work surface, which includes the whole area of the laminar airflow bench station should be swabbed with 70/90% ethyl alcohol. The working light and the airflow should then be switched on. The necessary tools are sterillized by overnight incubation of them in steam sterilizing units or dry sterilizing units such as ovens, which may reach a temperature of over 200°C. The usual tools, which are used for plant transfer, are forceps, scalpel, tweezers, etc. The scalpels should always be fitted with right sized surgical blade. The scalpels normally used are rated as 4 and 5. These scalpels are fitted with surgical blades, which are sized 10 and 22, respectively. Each blade should only be used to cut a definite number of explants. However, this depends on the species of plant and the kind of explant used. For example, the blade may not last longer or may turn blunt quickly with woody explant materials unlike material of herbaceous nature. Utmost care should be taken to fix the blade onto the scalpel. If the blade is not slipping onto the scalpel fairly easily, it should be checked for the compatibility of size of blade and scalpel. Any forcible fixing of the blade may cause accidental injury to the fingers or the hand or the blade may be broken. In the laminar flow, these tools may be sterilized regularly in between transfer of either individual explants or a batch of explants by inserting in steribead sterilizers. Dipping the tools for 30 to 40 seconds should be enough to render the tools sterile. These tools should not be left in the sterilizer for a prolonged time as it may cause damage to the quality of the blade. The sterile plant material should be placed on a sterile plate of glass or ceramic. In the absence of these, the students tend to use big sized glass Petriplates. The explant material is cut to ideal sizes by the surgical blades and transferred onto the media using long forceps. Using short forceps should be avoided, as hands during explant transfer may touch the neck of the bottle. Also, hands come closer to the mouth of the bottle, which may not be a good practice. Everytime an explant is cut and transferred onto the media, the basal portion of these tools should be rinsed in alcohol and then transferred to steribead sterilizers. By this time, another pair of sterile tools should be ready and the work of plant transfer should be carried out unhindered without wasting any time. If the transfer operation is carried out for subculturing, the culture bottle containing the material to be subcultured should always be cleaned sterile by spraying them with alcohol and then only they have to be brought into the laminar area. For efficient plant material transfer operations, it is important that the operator carries out the operations methodically. Clumsy operations always lead to problems and reduce the work efficiency. A good operator may transfer cultures to more than 500 bottles a day. The commercial laboratories are capable of accomplishing a couple of thousand of bottles a day. During the culture transfer operations, the excess or trimmed plant materials should be thrown into a waste-basket. Care should be taken to see that the hands are not stretched outside the laminar area. Hence, it is better to stick a polythene bag or a paper bag with the edge of the laminar flow cabinet and the left over plant material is pushed into the bag. If the transfer operations are accomplished for the day, the laminar area should again be swabbed with alcohol after clearing the other materials taken into the area. The lights and airflow should be switched off and the laminar covers, if any, should be lowered. It makes sense to say that the LAF is left as it was found before starting the culture operations.

5.2. CONTAMINATION OF PLANT TISSUE CULTURES

Contamination could occur at any time of an operation. It can occur by spore transfer or by contact with incompletely sterilized media, solutions and surfaces. Spores may be carried by air and when they germinate, the organism proliferates very quickly on the medium.

Contamination of the cultures may be due to bacteria, yeast or fungi. Mites also have been found to cause contamination of plant tissue cultures (Plate 15). Bacteria are recognized by a turbidity in liquid media, a growth with colouration and sometimes by unusual odours. Yeast growth often appears as a heavy 'milky' turbidity in liquid media and has a distinct odour, while fungi forms mycelia on the media surface.

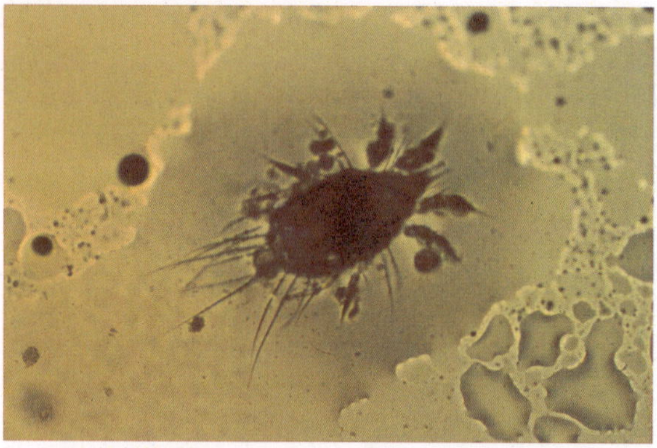

Plate 15: Contamination of plant tissue culture media by mites

Once a plant culture or media has been contaminated, the containers should be autoclaved, before the contents are discarded. Refer to Chapter 27 on Problems in Plant Tissue Culture for more information on culture contamination.

5.3. CONTROL OF CONTAMINATION BY THE USE OF ANTIBIOTICS

Antibiotics can be used to reduce, control or eliminate the microorganisms causing contamination in plant tissue culture. No known antibiotic is however, effective against all the microorganisms. It is our experience though antibiotics are sometimes used with great expectations but end up with insignificant results. Hence, a lot of restraint should be exercised while deciding the use of antibiotics in the media.

A good antibiotic should be soluble in the nutrient medium and there should be no significant change in the salt concentration due to its presence. The drug should be stable in the various adverse conditions it might encounter and must be unaffected by changes in pH and be compatible with the various media constituents. The drug should be systemic in plant tissue and should have no side effects on the tissue. Further, it should have a broad spectrum action and should be inexpensive. However, the antibiotics should never replace the aseptic techniques. Some of the commonly used antibiotics are as follows:

Table 7: Commonly used antibiotics and their mode of action

Antibiotic	*Mode of action*
Aminoglycosides Examples: Streptomycin Kanamycin	Inhibits protein synthesis by interacting with 30S or 50S ribosomes
Actams Examples: Penicillin Ampicillin Tetracyclins	Inhibits bacterial cell wall synthesis
Quinolones Examples: Nalidixic acid Enoxacin	Interferes with DNA replication
Polymixins Examples: Rifampicin	Inhibits bacterial membrane functions

5.4. RECOVERY OF CONTAMINATED CULTURES

- Fill the container containing the infested culture to the brim with a solution of 0.5-1.0% sodium hypochlorite
- Allow to stand in chlorine solution for 10-15 min depending on the severity of the infestation or the sensitivity of the plant material
- Remove the cultures from the chlorine solution, trim the base and remove excess leaves
- Transfer to fresh culture medium

It is optional whether the plants should be rinsed with sterile water or passed through a series of sodium hypochlorite solutions (1%, 0.5%, 0.25% and 0.01%) and inoculated onto the medium without a final rinse in sterile water. This means the piece of tissue finally placed on the sterile medium carries with it some chlorine. This is particularly useful in case of heavily contaminated cultures, however, only those plants, which are not sensitive to chlorine, can be treated thus. Also, this method of recovering infested cultures cause bleaching of the leaves. However, these cultures will soon recover and grow successfully.

If the infestatiom seen is less, infected explant culture could be saved by removing the explant/culture from the container and rinsing them in 90% Alcohol solution for a short duration 10-15 seconds. Holding or rinsing for a long duration can kill the explant/plant culture. After removing from Alcohol solution, the plant material is immediately transferred to sterile water and rinsed off the Alcohol and then could be transfered to fesh media.

Recovering or saving methods as explained here seldom work, but worth a try.

6

SAFETY RULES AND REGULATIONS

A person who is new to the plant tissue culture laboratory has to undergo certain orientation programmes to equip himself with the knowledge of the laboratory and the subject. It includes various instructions on how to operate equipment like pH meter, balance, laminar airflow cabinet, microscope and various other instruments. It is mandatory for any person working in a tissue culture laboratory to adhere to the following rules:

1. First aid kits should be kept in every laboratory and the staff should know its location and how to use its contents.
2. A fire extinguisher must be there in the laboratory.
3. Personal belongings like ornaments and footwear should be kept outside the laboratory or the working table.
4. Long hair must be tied up and nails must be trimmed to prevent accident or contamination.
5. Eating and drinking in the laboratory and leaning on the working table should be avoided.
6. An autoclaved or UV sterilized apron should be worn while working inside the laboratory.
7. Hands must be washed thoroughly before beginning aseptic work.
8. Toxic chemicals must be treated with precaution and discarded into separate labelled containers.
9. Do not work with UV light on. UV light should be put (if necessary) only for 15 minutes prior starting of any culture work. Before starting use of laminar flow cabinet, it should be confirmed whether the UV light is switched off. There are several incidences where either due to negligence or due to anxiousness to complete a work quickly, people tend to forget switching off the UV. Exposure of body parts to UV even for 15-30 minutes can cause severe irritation and itching of the skin. In worst cases, there is a danger of losing your eye sight. These have occured in laboratories. Though, research workers have not lost eyes, they certainly

have experienced severe itching and discomfort in the eyes needing doctor's assistance. So, extreme care is needed in this regard. Our experience tells that there is no need for UV light for tissue culture work.

10. The body and head of the person working in the laminar airflow chamber should be outside the hood.
11. The person sitting in front of the laminar airflow chamber should maintain a distance from the edge of the laminar airflow surface. Sitting closer touching the edge of the surface may contribute for microbial contamination as the laminar airflow might bounce back into the chamber touching the body of the person.
12. LAF cabinet must be put off and the HEPA filters must be covered once the work is over.
13. Maintenance of photoperiod, temperature and humidity as per the settings should be checked periodically.
14. The laboratory must be kept clean and microbe free environment should be maintained.
15. The person working must have his gloves on and a mask to avoid contamination.
16. Dispense with bunsen burners or any naked flame devises for instrument sterilisation. We have experienced students getting into serious problems of burn injuries and also contribute to fire hazard in the laboratories. The consequences could be disastrous.

An Averted Mishap

Here, I would like to narrate an accident which was unnerving. At my UK laboratory, one evening, Andy (a British student), Rabbani (a Bangladeshi student) and me were working infront of our laminar flow hoods. The evening was boring. Everybody was involved in his own work. Rabbani, looked uninterested to talk, looked tired and had a bored expression. Andy was not his usual jovial self. I settled down to my work. There was a palpable silence in the lab. Suddenly, Rabbani, threw a jar engulfed in fire down on to the floor, his palm was burnt. He was panick stricken and dumb founded. The floor vinyl caught fire. Ever (extra) cautious and agile Andy ran, to be pulled an extinguisher and calmed the fire. If Rabbani was alone?

The incident happened, as normally it is a practice to dip the instrument into a jar containing alcohol after cutting the plant material and transfering it into culture bottle and then flame it. Here, Rabbani (probably not concentrating) held an instrument over the flame and then dipped into the jar filled with Alcohol setting fire instantly. Another day, I did the same, my previous experience with Rabbanis episode saved me. Hence, it is advisible to use non flame sterlizers such as a steribead steriliser.

7

Applications of Plant Tissue Culture

Technology becomes essential to evaluate available resources with respect to the utilitarian value, make them available on a sustainable basis for future generations and convert them into products of economic value. Tissue culture is a hugely successful viable technology for the multiplication of superior plants on large-scale on a perpetual basis, production of novel and improved plants and for biosynthesis of products of industrial and medicinal value from natural plant resources. Tissue culture in simple terms refers to the cultivation of the cells and tissues free from the mother plant in appropriate nutrient substrates under sterile conditions. The cells in culture divide, multiply and produce compounds specific to the plants and also differentiate. Since these plants are identical to modern plants, tissue culture has been effectively used for commercial production and delivery of plants by private enterprises. In the world, there are more than 800 companies involved in the business of plant tissue culture. More than 50 biotech industries are producing roughly 110 million plants annually for the export market. This is out of an estimated production of 850 million plants worldwide with an annual growth rate of 19%. Net profits, multinational companies aim to fetch via production of pharmaceuticals that could be extracted from plant resources, is in the order of US$76 billion.

Applications of plant tissue culture will be particularly rewarding in the areas of agriculture, horticulture, floriculture, agroforestry, plant medicine and gene transfer across species and families. Tissue culture and DNA technology together make what we call plant biotechnology in modern times.

Plant tissue culture now has direct commercial applications as well as value in basic research into cell biology, genetics and biochemistry. The techniques include culture of cells, anthers, ovules and embryos on experimental to industrial scales, protoplast isolation and fusion, cell selection and meristem and bud culture.

Applications of Plant Tissue Culture

- Virus elimination
- Rapid Multiplication
- Micropropagation
- Crop Improvement
- *In vitro* germplasm conservation
- Secondary metabolite production

Table 8 lists the main applications of plant tissue culture.

7.1. MICROPROPAGATION

These methods could be employed for either elimination of viruses form infested plants or for rapid multiplication.

Incubation of plants or cultures at relatively high temperatures (32° to 40°C) either alone or in combination with the excision and culture of very small stem tips or meristems is the only method of ridding infected cultivars of their viruses.

Ideally, it is advised to excise and culture the apical meristem of an infected plant because this region is most likely to be free of viruses. In fact, this would involve a very delicate dissection and such small explants are difficult to grow in culture. Instead, we excise a very small stem tip, which in addition to the apical meristem, has one or two leaf primordia. These are much easier to grow in culture and, although a high magnification stereomicroscope is still needed, their dissection is not so difficult.

However, some viruses invade the apical meristem and, in these cases, it is necessary to apply heat treatments to the plant. In the past, this has meant heat treatment of the whole plant followed by stem tip culture. An easier method involves the application of the heat treatment to aseptic cultures.

A generalized protocol is as follows:

1. Disinfest plant material and culture buds.
2. Discard contaminated cultures and rear aseptic cultures until they are small plants.
3. Expose these aseptic plants to elevated temperatures during incubation for various times (to be standardized).
4. Excise small stem tips from these incubated plants and incubate in culture at 25°C.
5. Incubate until each plant is quite large and test for presence of viruses.

6. If free of detectable viruses, re-culture from this (or these) plant(s) and multiply to obtain a population of virus-free clonal material.

Micropropagation is defined as the plant propagation of selected genotypes using *in vitro* culture techniques. Depending on the species and cultural conditions, *in vitro* propagation can be achieved by the following four basic methods:

- Enhanced axillary shoot proliferation (Shoot culture),
- Node culture,
- *De novo* formation of adventitious shoot through shoot organogenesis, and
- Somatic embryogenesis.

Currently, the most frequently used micropropagation method for commercial production utilizes enhanced axillary shoot proliferation from cultured meristems. This method provides genetic stability and is easily attainable for many plant species. Getting an organized structure such as a shoot into aseptic culture is a time consuming process and, in general, the earliest stages of micropropagation involve loss of cultures through either microbial contamination or death of cultures. Multiplication rates also tend to be slow at first but later, if the cultural conditions are satisfactory, a rapid multiplication can be achieved.

Under appropriate cultural conditions, if a cultured bud develops a shoot with 5 or more nodes over a 2-month period of incubation; each of these nodes can then be excised and cultured on fresh medium and, after a further 2-month period of incubation, each subcultured node would develop a shoot with 5 or more nodes. Thus, we would have a multiplication rate of × 5 every 2 months, which would give us $5 \times 5 \times 5 \times 5 \times 5 \times 5$ ($= 5^6 = 15{,}625$) nodes for further culture or for establishment, after 12 months.

Another type of multiplication that has been attained involves the inhibition of apical dominance of the cultured bud. When we prune a shrub, we remove the growing points, (i.e., apices) of the stem and branches, and thus remove their apical dominance (i.e., their inhibitory function) over the buds in the axils of leaves on the remaining parts of the stem and branches. These no longer inhibit the development of buds into shoots and branches. In organ culture, we have been able with some species to overcome apical dominance by chemical means. Thus, under appropriate cultural conditions, the initial bud cultured grows only slightly before developing lots of axillary buds, which are separated from each other by only very short internodes. Each of these axillary branches grows slightly and then develops its axillary buds, and so on. Often the resulting cultures bear little resemblance to the normal plant and take on a grass like or moss like habit. Later, these buds are planted on different media or are exposed to different incubation conditions to restore them to their normal habit.

Care has to be exercised with this type of multiplication to ensure that a reversal of habit in culture is possible; cultures exposed to this type of multiplication system are also prone to initiate adventitious bud formation in addition to axillary buds. This type of multiplication system is capable of producing in excess of 10 buds during a 2-month incubation period, that is $10 \times 10 \times 10 \times 10 \times 10 \times 10$ (= 10^6 = 1,000,000) buds over a 12 month period.

7.2. CROP IMPROVEMENT

Important applications of tissue culture in crop improvement are the production of new mutants, genomic recombination and selection of new recombinants by:

- Anther culture for haploid production
- Embryo/ovule culture after interspecific hybridization
- Genetic engineering (Transformation)
- *In vitro* selection
- *In vitro* germplasm conservation and exchange
- Cell and organ culture
- Somaclonal variation
- Somatic cell hybridization (protoplast fusion)
- Somatic embryogenesis

7.2.1. Anther and Pollen Culture

The culture of immature anthers and pollen grains is done so as to induce the pollen grains to develop into multicellular forms, particularly into embryos, with half the normal number of chromosomes for the species. When such haploid embryos are treated with chromosome doubling agents, e.g. colchicine, their normal chromosome number is restored (and thus their fertility) and the achieved plants are pure lines. Pure line formation is a natural tendency for self-fertilizing species and can be obtained with cross-fertilizing species only after repeated inbreeding for 10 or more generations.

So far, the induction of haploid plant formation from anther and pollen cultures has been successful mainly with naturally self-fertilizing species and thus, on chromosome doubling, are in theory very similar if not identical with the parents. However, by first crossing many lines from a self-fertilizing species, new combinations of genes are formed, and haploid plants produced by anther culture from such crosses can be an extremely valuable and quick way of obtaining pure lines of these new combinations.

7.2.2. Callus and Suspension Cultures

It is relatively easy to induce callus and suspension cultures from many species, including woody species. So far, we have been able to induce organogenesis and/or embryogenesis with very few species. Polyploid cells and cells with various types of chromosomal variations from the normal for the species have often been reported from callus and suspension cultures, and thus plants which are induced to form from such cultures, either by the induction of adventitious buds or by embryogenesis, may arise from cells of an abnormal type. Nevertheless, the presence of cells with other than normal genomes does not necessarily prevent the formation of adventitious buds and embryos with the normal chromosomal type of the species. On the other hand, this variation in genome could lead to the formation of interesting and valuable new strains. However, for clonal propagation, we have at present to use the callus and suspension culture methods with extreme caution. But for somaclonal variation it might be of considerable importance in plant breeding. The major requirements for using somaclonal variation in plant breeding are:

- Techniques to culture tissues and regenerate genetically stable plants
- Techniques to screen regenerated plants for desired variants

7.2.3. Protoplast Culture

Enzymatic removal of the mainly cellulose wall surrounding a plant cell results in a protoplast. When these are cultured, a new wall is synthesized thus restoring the original state of the plant cell. However, during the period that the protoplast is wall-less, we have the possibility to:

1. Add foreign genetic material into the protoplast,
2. Fuse protoplasts of two cultivars or even species to form hybrid cells.

The latter is called somatic hybridization and can be used to overcome problems related to sexual incompatibility experienced within different species.

7.2.4. Transformation

Plant transformation (or genetic engineering) involves the transfer of specific foreign DNA into a plant species resulting in a transgenic plant, *i.e.*, a plant that has been transformed. Transformation involves:

1. Isolation of a useful gene
2. Transfer of the gene into plant cells
3. Integration of the gene into the plant genome
4. Regeneration of fertile plants
5. Expression of the transgene in the regenerated plants
6. Transmission of the transgene from generation to generation

7.2.4.1. Some Methods for Transferring Foreign DNA into Plant Cells

- Use of *Agrobacterium tumefaciens* as a vector

- Direct DNA uptake—facilitated by polyethylene glycol or electroporation
- Microprojection of DNA into cells with a particle gun
- Microinjection
- Silicon carbide crystals

7.2.4.1.1. *Agrobacterium-mediated Transformation*

Agrobacterium tumefaciens is a gram negative aerobic bacterium which occurs as a free-living soil bacterium. It can infect plant tissues through open wounds. Upon infection, dicots respond to infection by initiating massive cell division at the point of infection. Following the initiation of cell division, plant cells near the site of infection begin production of unusual amino acids (opines). Plant cells are unable to catabolize these unusual metabolites, but *Agrobacterium* cells can catabolize them, using them as both carbon and nitrogen sources.

There are at least three species of *Agrobacterium: A. tumefaciens, A. rhizogenes*, and *A. radiobacter* capable of infecting plants. *A. tumefaciens* initiates gall formation, generally at around the soil surface near the crown. Hence the name 'crown gall' for the tumor-like structure that forms following infection. Through infection and gene transfer, *A. tumefaciens* accomplishes two objectives. First, it obtains a protected environment in which to grow. Second, it engineers the plant cells surrounding its colony to produce unusual amino acids which the plant cell cannot catabolize, but which the bacterium can. The bacterium creates for itself a marvellous, nutrient-rich home. From the evolutionary perspective, the genes, which *Agrobacterium* introduces, have never been observed to be sexually transmitted in nature.

The infection process is composed of:

- Chemotaxis and initial binding at the site of wounding
- Development of a stable attachment
- Initiation of t-DNA excision and packaging
- Construction of transmembrane channel
- Movement of t-DNA to plant cell
- Movement of t-DNA to nucleus, integration and transcription

The Ti plasmid present in the *Agrobacterium tumefaciens* carries two significant gene regions, the *vir* region and the t-DNA. The t-DNA is the DNA that is transferred. Most of the genes required for infection are carried within the *vir* region.

In the *vir* region of the plasmid are the *vir* A, *vir* B, *vir* C, *vir* D, *vir* E, *vir* G and *Vir* H operons. As an example of the internal complexity of these, *vir* B contains 11 characterized open reading frames. *Vir* A produces the protein most directly associated with chemotaxis and sensing of plant phenolic compounds (notably acetosyringone) associated with wounding. *Vir* H may act as a detoxifier of the plant bacteriocidal compounds produced through the wound response.

7.3. *IN VITRO* GERMPLASM CONSERVATION

Germplasm, as defined for a crop, represents all the different genotypes that could be used for improvement. Germplasm diversity for any crop must be protected from loss to ensure its availability for future improvement. The *in vitro* germplasm resources conservation include:

- Cultivated species
- Commercial varieties
- Breeding lines
- Landraces
- Special genetic stocks
- Wild species
  - for direct use—currently used or potentially useful as crop species
  - for indirect use—in crosses, as rootstocks

With conventional breeding techniques, the interest was limited to species that could cross with cultivated species. But now with the modern genetic engineering techniques, it has become essential to conserve all species.

7.4. SECONDARY METABOLITE PRODUCTION

Of late, there has been considerable interest in investigating the potential of plant cell cultures as an alternative to the traditional method for the industrial production of secondary plant metabolites. Cell suspension systems could be used for the large-scale culturing of plant cells from which secondary metabolites can be extracted. Despite technical difficulties, more than 30 natural products have been biosynthesized and accumulated in plant cell and organ cultures at levels higher than those produced in the corresponding whole plants. Manipulation of the culture environment and media can affect the rates of both cell growth and the accumulation of secondary metabolites.

Facilities for the large-scale production of secondary metabolites from plant cells have been developed, but further research is needed to improve their economic efficiency. In the future, valuable pharmaceutical intermediates and drugs could be produced on an industrial scale in bioreactors. However, a large part of the success will rely on the ability to obtain high yielding and genetically stable cell lines with shorter doubling times on refinement of contamination free culture methods and on improvements in bioreactor design. Most of the plants taken up for secondary metabolite production are at present drug yielding herbs.

Table 8: Applications of plant tissue culture

Tissue Culture Technique	*Applicaitons*
Seed Culture	• Increasing efficiency of germination and seedling production in seeds, which are difficult to germinate *in vivo* • Precocious germination by application of plant growth regulators • Elimination of viruses as seeds does not carry viruses
Embryo Culture	• Overcoming embryo abortion due to incompatibility barriers • Overcoming seed dormancy and self-sterility of seeds • Embryo rescue in distant (interspecific or intergeneric) hybridization where endosperm development is poor • Production of monoploids • Shortening of breeding cycle • For development of callus cultures
Ovary or Ovule Culture	• Production of haploid plants • Recovery of hybrid embryos by overcoming embryo abortion at very early stages of development of zygote due to incompatibility barriers • Achievement of *In vitro* fertilization
Anther and Microspore Culture	• Production of haploid plants • Production of homozygous diploid lines through chromosome doubling, thus reducing the breeding cycle • Genetic transformation using microspores. • Production of useful gametoclonal variations • Mutation investigations easier with single set of chromosomes • Fixation of certain genetic characters from heterozygous source materials
In vitro Pollination	• Production of hybrids difficult to produce by embryo rescue
In vitro Fertilization	• Production of distant hybrids avoiding style and stigmatic incompatibility that inhibits pollen germination and pollen tube growth • Production of transgenics by injecting exogenous DNA in the nuclei of gametes and zygotes
Somatic Embryogenesis	• Mass multiplication of elite germplasm • Production of artificial seeds • As source material for embryogenic protoplasts • For genetic transformation • Production of primary metabolites specific to seeds such as lipids in oil seeds • Amenable to mechanization and for bioreactors
Organ Culture	• Mass production of plants of elite and rare germplasm

Table 8: Continued

Tissue Culture Technique	*Applicaitons*
	• Production of calli, shoots and roots for production of secondary metabolites • Development of germplasm banks for rare and endangered plants
Organogenesis and Enhanced Axillary Budding	• Mass multiplication of elite germplasm • As source material for protoplast work, genetic transformation and micrografting • Conservation of endangered genotypes either at normal or at sub-zero temperatures
Callus Cultures	• Production of plantlets through somatic embryogenesis or organogenesis • For obtaining virus-free plants • For generation of useful somaclonal and gametoclonal variants • As a source of protoplasts and suspension cultures • Production of useful secondary metabolites • For biotransformation studies • Selection of cell lines with valuable properties such as resistance to disease, herbicides, overproduction of secondary metabolites etc. • For mutagenetic studies
In vitro Production of Secondary Metabolites	• Production of useful compounds such as drugs, aromatic substances, pigments, flavours etc. without destruction of mother plants • Production of novel metabolites normally not produced by the parent plant • Biotransformation and elicitor studies
Cell Culture and In vitro Selection at Cellular Level	• Production of somatic embryos, morphogenetic nodules and entire plantlets • Over-production of secondary metabolites • Over-production of primary metabolites • Induction and selection of useful mutants or somaclones at cell level for disease resistance, stress tolerance and improved nutritional quality in less time and space
Somaclonal Variations (Genetic or Epigenetic)	• Isolation of useful variants in well-adapted, high yielding genotypes lacking in a few desirable traits • Isolation of useful variants overproducing primary or secondary metabolites • Isolation of useful variants with better disease resistance, stress tolerance capacities • Creation of additional genetic variation without hybridization in useful cultivars
In vitro Mutagenesis	• Induction of polyploidy for consequent increase in biomass or yield

Table 8: Continued

Tissue Culture Technique	*Applicaitons*
	• Introduction of genetic variability and rapid selection as well as multiplication of useful mutants • As a tool for developmental genetics and for elucidation of biochemical processes
Protoplast Isolation, Culture and Fusion	• Combining distant genomes to produce somatic hybrids, asymmetric hybrids and cybrids • Production of organelle recombinants • Transfer of CMS (cytoplasmic male sterility) in elite lines • Source material for genetic transformation • Creation of genetic variants
Genetic Transformation	• Introduction of foreign DNA to generate novel genetic combinations • Transfer of desirable genes for disease and pest resistance from related or unrelated plant species into high yielding susceptible cultivars • Study of structure and function of genes • Induction of hairy roots or shooty teratomas for over-production of secondary metabolites, naturally present in mother plant • Production of novel secondary metabolites absent in parent plant
In vitro Flowering	• Reduction in long life cycle in perennials such as Bamboo • Continuous supply of flowers, fruits and seeds irrespective of season
Micrografting	• Overcoming graft incompatibility • Rapid mass propagation of elite scions grafted on rootstocks having desirable traits like resistance to soil-borne pathogens and diseases (Plate 16) • Multiplication and survival of difficult to root species as well as of transformants • Development of virus-free plants
Cryopreservation or Storage at Low Temperature	• Long term preservation of useful germplasm (cell lines, meristems, plant organs, morphogenetic callus cultures) • Conservation of natural genetic variability
Culture of Protoplasts, Cells, Tissues and Organs	• As a tool in Phytopathological Research • Virus preparation and replication • Culture of obligate parasites • Host-parasite interactions • Culture of nematodes (Excised root cultures) • Testing of phytoalexins and phytotoxins • Nodulation studies • As a tool in Plant Physiological Research

Table 8: Continued

Tissue Culture Technique	*Applicaitons*
Culture of Hairy Roots	• Cell cycle studies • Metabolic studies • Nutritional studies • Morphogenetic and developmental studies • For understanding and manipulating root-specific metabolism • For co-culture with VAM fungi to increase secondary metabolite production • For co-culture with insects to study pathogenesis • For study and commercial exploitation of bioactive root exudates • For co-culture with shooty teratomas to exploit both root and shoot based metabolism for biotransformations and also for metabolite production specific with both these locations • For development of "green hairy roots" which are photoautotrophic and hence display a different spectrum of secondary metabolites

Plate 16: Micro-grafting of a clove scion over guava stock

8

MICROPROPAGATION

Micropropagation has been defined as "*in vitro* regeneration of plants from organs, tissues, cells or protoplasts" (Beversdorf, 1990). Micropropagation is a method of propagating plants using very small parts of plants and are cultured in sterile conditions. Micropropagation, as an application of tissue culture is the most exploited facet commercially. Though this application may not be the most sought after by people involved in plant breeding, but certainly serves as a pre-requisite for any crop improvement programmes. For example, a genetically altered explant has to be regenerated into producing new plants. Otherwise, the objective of genetic alteration serves no purpose. Hence, any scientist desirous of crop improvement through genetic manipulations should first embark on developing a feasible protocol for regeneration *in vitro* of the species.

George Morel (1960) first showed the potential of clonal propagation in culture by describing *in vitro* multiplication of *Cymbidium* orchids. Small, spherical bodies, called protocorms, formed on cultured shoot tip explants. Several protocorms formed on each explant, and when divided and transferred, additional protocorms were formed. The rate of multiplication was such that several million plants could be produced from one shoot tip in a single year. As a result, many of the economically important orchids are now usually propagated *in vitro*, although techniques have not been worked out for all orchid types.

However, credit for the development of the field of micropropagation in general goes without a doubt to Murashige, who showed that many plants, in addition to orchid, could be propagated *in vitro*. Micropropagation has been shown to be especially useful in horticultural crops.

The methods that are theoretically available for the propagation of plants *in vitro* are essentially:

(*a*) Multiplication of plants by the release and growth of pre-existing (axillary or lateral) meristems in the initial culture; this is often referred to as shoot culture. Some scientists describe this classic method as axillary bud proliferation.

(*b*) Multiplication of plants by the formation of

1. Adventitious shoots and/or
2. Adventitious somatic embryos
   - Directly on pieces of tissue or organs removed from the mother plant
   - Indirectly on unorganized callus tissue or in suspension cultures
   - Indirectly on semi-organised callus tissue or propagation bodies (protocorms or pseudo-bulbils) obtained from explants.

The production of plants from axillary shoots has proved to be the most generally applicable and reliable method of *in vitro* propagation. Theoretically, any part of a plant could be used for initiating *in vitro* cultures (Fig. 2).

Fig. 2: Plant parts used as explants for micropropagation

8.1. MICROPROPAGATION BY THE RELEASE AND GROWTH OF PRE-EXISTING MERISTEMS

Shoot meristems

In plants, organ initiation and development continue throughout the life cycle. This is in contrast to animals in which all organs are laid down during embryogenesis. Furthermore, there is plasticity in plant development that allows for development according to environmental conditions. Since all plant organs develop from meristems, these structures are where plant development and plant form ultimately is regulated. Meristems are formed during embryogenesis and consist of groups of undifferentiated cells that will initiate organ primordia during plant life. The apical shoot meristem give rises to the complete shoot of the plant and the apical root meristem is the origin of the root.

Different zones can be distinguished within the apical shoot meristem (Plate 17).

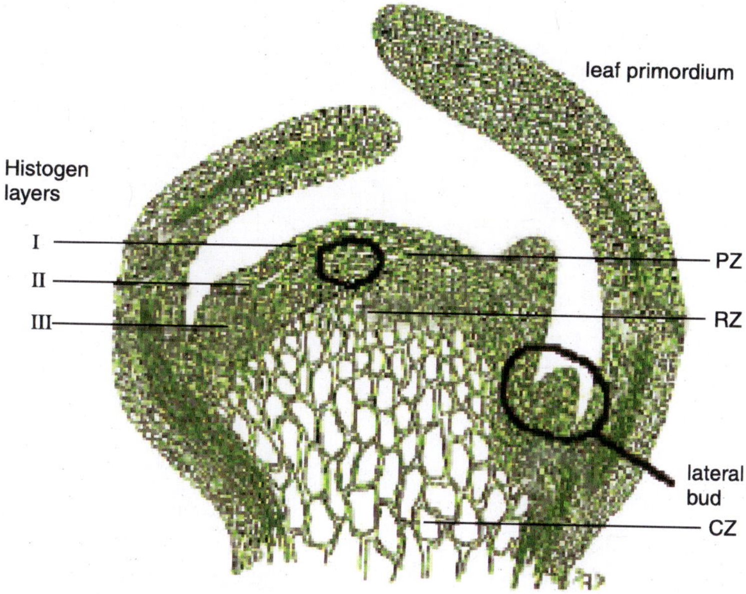

Plate 17: Cross-section of shoot apical meristem

Cells in the central zone (CZ) are large and divide infrequently while the peripheral zone (PZ), the site of organogenesis, consists of small cells that divide fast and differentiate. The third zone is the rib zone (RZ) giving rise to cells of the stem and forming the boundary against fully differentiated cells. Like cells in the peripheral zone the rib cells originate from the undifferentiated cells in the central zone. Apart from forming cells of the stem, the rib zone has been suggested to act as an organizing centre for the shoot. In vegetative and inflorescence meristems, the central cells play a role comparable to animal stem cells as they are characterized by their undifferentiated state and their ability to regenerate themselves as they give rise to daughter cells differentiating into specialized cell types.

In stems, mitosis in the apical meristem of the shoot apex (also called the terminal bud) produces cells that enable the stem to grow longer and, periodically, cells that will give rise to leaves. The point on the stem where leaves develop is called a **node**. The region between a pair of adjacent nodes is called the **internode**.

Shoot apical meristems are found at the terminals (tops) of shoots. Further down the shoot are axillary, or lateral meristems. There is typically an axillary bud in each leaf axil on the shoot and each encloses a shoot tip (Plate 18). Each of these buds has the potential to become a shoot under the proper conditions.

In vivo, the growth and development of axillary buds is frequently suppressed by the apical dominance of the apical meristem, which produces auxin. *In vitro*, cytokinins can be added to the medium to overcome apical

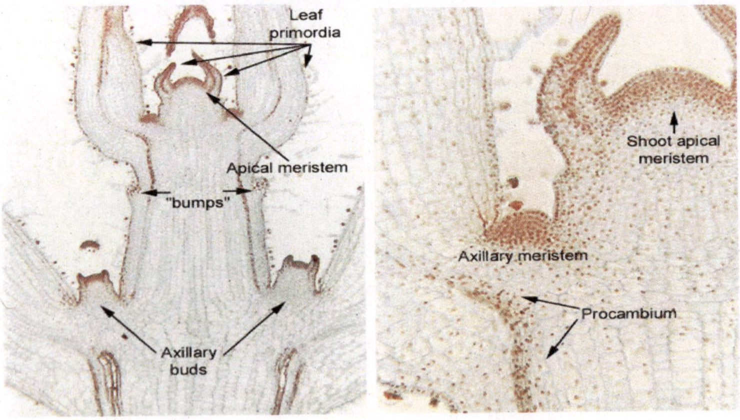

Plate 18: Cross-section of shoot tip

dominance and release the growth of the axillary meristems. Each axillary branch can also be excised and transferred to a fresh appropriate medium for enhanced axillary branching. The branches may be later transferred to a rooting medium and subsequently transferred to soil.

Examples: Most herbaceous ornamental species, potato, most woody species.

8.2. MICROPROPAGATION BY ADVENTITIOUS SHOOT PROLIFERATION

In this method, shoots arise directly from plant parts or tissues other than from preformed meristems. Direct shoot formation may be accompanied by some proliferation of unorganized cells that may be termed as callus and some cells of the callus tissue group together to regenerate organs. Proliferative tissue of this kind ideally should not be used for new subcultures. The induction of direct shoot regeneration depends on the plant organ from which the explant was derived and above all, on the plant species. Rates of propagation can be quite high by this method, particularly if numerous small shoots arise rapidly from each explant. The second author, working on *Bacopa monnieri*, a medicinal plant observed the direct emergence of numerous plantlets from leaf and internodal explants cultured on MS basal media without any growth regulators. The plantlets, when maintained on the same media produced roots also, hence enabling the harvest of rooted plantlets, which could be directly acclimatized.

Examples: *Achimenes, Saintpaulia, Streptocarpus, Begonias, Gerbera, Hosta, Lilium, Bacopa monnieri, Pogostemon patchouli*.

8.3. PLANT REGENERATION BY DIRECT SOMATIC EMBRYOGENESIS

Direct somatic embryogenesis is the formation of somatic embryos or embryogenic tissues directly from the explant without the formation of an intermediate callus phase. To induce and study direct embryogenesis, the explant most often used is the immature zygotic embryo. This tissue is already embryogenic in nature and apparently requires less nurturing than other somatic tissues to induce a somatic embryogenic response. However, direct somatic embryo production may not auger well for large-scale clonal propagation as the number of direct embryos, obtained would be less per explant. When ovules, nucellar embryos or nucellus tissues are excised from plants capable of nucellar embryony and cultured *in vitro*, there can be an enhanced likelihood of direct adventive embryo formation.

Direct embryogenesis, as said earlier, is not a common occurrence. To induce direct somatic embryogenesis from explants, especially other than the plant parts which have predetermined meristems, it may be useful to use BAP in the media. If auxin is needed, it should be included along with BAP especially at a ratio of higher cytokinin (BAP) and lower auxins. Direct somatic embryogenesis has been reported in carambola (star fruit) from *in vitro* grown seedling roots as explants on a medium containing BAP (Sathyanarayana *et al.*, 1995).

Examples: *Asplenium, Cardamine, Kalanchoe*, Citrus, Carambola.

8.4. PLANT REGENERATION BY INDIRECT SOMATIC EMBRYOGENESIS

Indirect formation of somatic embryos from callus or suspension cultures is observed more frequently than direct somatic embryogenesis. The process of successful somatic embryogenesis has the following requirements:

- For the initiation of embryogenic cells or somatic embryos, cultures should be grown in the presence of an auxin.
- After initiation of embryogenesis, it is usually necessary for the embryogenic tissues to be subcultured to a medium containing reduced auxin concentration or no auxin at all. It is often necessary in some species to transfer the embryogenic cultures to be placed on medium containing Abscisic acid to aid maturation of embryos.

Mass propagation by somatic embryogenesis can be best-achieved using suspension cultures. Horticultural crops like carrot have become regular and useful systems for indirect somatic embryogenesis studies due to the relative ease with which carrot respond to culture *in vitro*.

Examples: Carrot, Alfalfa, Oil palm, *Antirrhinum majus* and many plant species.

8.5. APPLICATIONS OF MICROPROPAGATION

- Clonal mass propagation of large numbers of uniform plants at a rapid rate.
- Micropropagation may allow faster production of plants that are slow to propagate *in vivo* in some plant species.
- It may decrease the time needed for bulk-up of new cultivars before they are introduced commercially.
- Disease free plants could be obtained as the process of micropropagation is done under aseptic conditions. With the use of meristem culture, virus-free plants could also be obtained.
- It is possible to obtain year round production as seasonal changes *in vivo* are less likely to affect cultures *in vitro* and that the conditions *in vitro* are artificially controlled.
- In case of plants that are threatened with extinction tissue culture propagation appears to be only hope for the survival and spread of the species.
- In case of plants that are difficult to be propagated by conventional methods such as grafting, budding, layering and cutting, tissue culture propagation offers a chance.

8.6. STAGES OF MICROPROPAGATION

Generally, when micropropagation was conceived as a fantastic application of plant tissue culture, Murashige (1974) pursuing further describes three basic stages such as Development of aseptic cultures (Stage I), Proliferation of shoot cultures also referred to as multiple shoot induction (Stage II) and Rooting *in vitro* (Stage III). However, Debergh and Maene (1981) have recognized five different stages of micropropagation. In that, they emphasized that a stage describing the selection of mother plant as the most crucial stage. They also recognized a stage of weaning *in vivo* as most important. In the stage III, which is described as the rooting stage, a sub-stage called IIIa is assigned to a stage of elongation of microshoots. This is in recognition of the fact that from many plant species microshoots, which are developed, are quite short and are difficult to be induced to root. Even if these shorter microshoots root, they do not make a good plantlet to be transferred to hardening stage. The stages recognized by Debergh and Maene is as follows:

Stage 0: Preparative Stage: Donor Plant Selection and Preparation

It is essential to give careful attention to the selection of a mother plant, which is typical of the variety and disease free. Growth, morphogenesis and rates of propagation *in vitro* can be improved by appropriate environmental and chemical pre-treatment of the mother plants. It may also be advantageous to treat the plant (or parts of it) in some way to make the *in vitro* culture successful. Explant quality and responsiveness is influenced the physiological phytosanitary condition of the donor plants. Important points to be noted are:

- Donor (stock) plants indexed for pathogens.
- Pathogen-free stock plants maintained in clean conditions (low humidity, drip irrigation).
- Vigorous growth is encouraged, but not over-fertilization.

Stage I: Establishment of Aseptic Culture

The initial step in the micropropagation process is to obtain an aseptic culture of the selected plant material. Success at this stage requires that explants should be safely transferred to the culture environment and there should be an appropriate response like growth of a shoot tip or formation of callus. This stage involves:

- Surface-sterilization / disinfestation: Must free explant tissues of all contaminating microorganisms, but not cause phytotoxicity.
- Isolation of shoot tip under sterile conditions.
- Medium—Must contain all components necessary to nourish explant and to make the explant perform as desired.
- Environmental conditions like light, temperature and relative humidity must be controlled.

Stage II: Production of Suitable Propagules

The objective of stage II is to bring about the multiplication of organs and structures that are able to give rise to intact plants through the formation of axillary shoots, adventitious shoots, embryoids, or miniature storage or propagative organs. Shoots produced at this stage can be thought of as propagules in that they can usually be cultured again to increase their number. Shoots harvested and shoot clusters transferred to fresh medium at frequent, regular intervals. Repeated enhanced axillary shoot production is encouraged by cytokinin in the medium, alone or with a smaller amount of auxin. Amount of cytokinin and presence and amount of auxin must be determined empirically. Number of subcultures possible from the original culture varies with species/ cultivar.

Stage III: Preparation for Growth in the Natural Environment (Shoot Elongation and *in vitro* Rooting of Shoots)

This stage involves the *in vitro* rooting of the shoots prior to their transfer to soil. Although some species form adventitious roots on shoots formed *in vitro* during the course of stage II culture, for many others, it is first necessary to grow elongated shoots ready for rooting and then utilise special media or methods to induce root formation. Alternatively, the shoots may be rooted *ex vitro*. (Refer Chapter 24 on *In vitro* and *ex vitro* rooting of microshoots for details). This method reduces costs and produces roots which are structurally and physiologically better. The damage to roots during transplanting can be avoided. Economically, it makes sense to follow *ex vitro* rooting methodology if the plant species is responsive. Following this stage (*ex vitro* rooting), the cost on production at *in vitro* rooting stage could be saved. *Ex vitro* rooting and hardening are accomplished simultaneously.

Stage IV: Transfer to the Natural Environment

The transfer of plantlets from the *in vitro* to the *ex vitro* environment is extremely important. If not carried out carefully, this transfer can result in a significant loss of plants. This process is referred to as acclimatization. Acclimatization is the process by which the plantlets physiologically and anatomically adjust from *in vitro* to *ex vitro* conditions. This is a relatively slow process and may take weeks. Acclimatization is important for better survival of the plantlets because the plantlets:

- Must adjust from high to lower relative humidity (e.g., from 98-99% to 20-60%) which requires the development of sufficient defences to control water loss.
- Epicuticular wax needs to be formed.
- The *in vitro* plantlets show abnormal stomatal development and function.
  a. Either are not properly depressed.
  b. Or Stomatal movement *i.e.*, opening and closing do not occur properly. In most cases the stomata do not close thus allowing loss of moisture and accounting for dehydration and irreversible collapse of plantlets.
- They also have non- or poorly functional roots.
- Should adjust from clean vs. presence of pathogens.
- Must adjust from low light to high light: from low photosynthetic competence to high photosynthetic competence. This is required due to the following characteristics of the *in vitro* plants.
- Poorly differentiated leaf structure.
- Poorly developed chloroplasts.
- The *in vitro* plantlets are heterotrophic and upon transfer, they soon become autotrophic in order to fix carbon independently.

(More information on acclimatization is given in Chapter 23)

9

Micropropagation by Proliferation of Axillary Buds

Micropropagation is the propagation of selected plant lines through tissue culture. The simplest type of *in vitro* plant propagation is the stimulation of axillary bud development. This technique exploits the normal ontogenic route for branch development by lateral (axillary) meristems (preformed meristem). The axillary buds are treated with hormones to break the dormancy and produce new shoots. The shoots are then separated and rooted to produce plants. Alternatively, the shoots can be further multiplied before rooting.

The plants can be produced from axillary shoots by following two methods:

- Shoot tip culture
- Single node culture

9.1. SHOOT TIP CULTURE

Shoot tip culture is by far the most commonly used *in vitro* method for commercial micropropagation. The explant used to start cultures of this kind is a lateral or main shoot apex, which may be up to 20mm in length. Dominance of apical meristems is removed and the development of precocious axillary shoots is stimulated by incorporation of growth regulators (usually cytokinins) into the growth medium.

9.2. SINGLE NODE CULTURE

Single node culture is another technique that can be used for propagating plants from axillary buds under *in vitro* conditions. In this method, shoot tip explants are grown into unbranched shoots of up to 5-10 cm in length bearing several discrete and separate nodes. Instead of inducing axillary shoot growth with growth regulators the following manipulative methods may be used to overcome the apical dominance and to promote lateral bud burst:

- The shoots may be placed on fresh medium in a horizontal position. This method is termed as '*in vitro* layering' and has been used to propagate potatoes *in vitro*.
- Each shoot may be cut into single node pieces, which are subcultured.

Single node culture is of importance for propagating species that produce elongated shoots in culture (e.g., Potato, *Alstroemeria*) especially if stimulation of lateral bud initiation are difficult to bring about with available cytokinins.

Objectives

- ➢ To initiate aseptic cultures of the plant provided
- ➢ To observe the formation of shoot branches from the axillary buds

Equipments

- Laminar flow hood
- Forceps, scalpels
- Sterile beakers, petri dishes

Media and Reagents

- MS media supplemented with BAP (0, 1, 2, 3, 4 and 5 mg/L)
- MS media supplemented with NAA (1mg/L)
- Distilled water
- 70% alcohol
- Sodium hypochlorite/Mercuric chloride
- Agar (0.8%)
- Sterile water

Plant Materials

Obtain axillary shoots from healthy plants free of disease symptoms or pest problems.

Note: Aseptically germinated seedlings are an ideal source of shoot explants in case of species where the germination rates are quite high.

Procedure

(*a*) Explant Collection and Preparation

- Explants from healthy, vigorous plants are more likely to produce successful cultures.
- Shoot tips and most recent flushes of growth are best. Avoid material in contact with the soil where the infestation is likely.
- If plants exhibit lateral branches or multiple stems, excise them for explant preparation. If no lateral branches are present, excise the upper 10-15 cm of the shoot.

- Soak the shoots in a solution of liquid detergent for 10 minutes, then rinse in running tap water for 5 minutes.
- To avoid possible contamination by fungi, soak the shoots in a fungicidal solution for 30-45 minutes.
- Rinse the shoots in distilled water, trim the leaves and reduce the size of the shoots to about 5 cm.

(*b*) *Surface Sterilization*

- Treat the explants with sodium hypochlorite/mercuric chloride solution or any other surface sterilent inside the laminar airflow chamber for the required time. The concentration of the chemical and the time of treatment depend on the plant selected.
- Alternatively, other surface sterilents can also be used.
- Addition of a surfactant allows easy penetration of the sterilent.
- After surface sterilization, rinse the explants in several changes of sterile water to remove traces of the chemicals present.

(*c*) *Culture Initiation*

- Trim the explants to the required size and place the prepared explants on to the media (MS media supplemented with BAP at different concentrations). If culture bottles are used, 3 to 4 explants can be placed in a bottle.
- Seal the culture vessels and incubate them for 4 weeks.

(*d*) *Multiplication of the Shoots*

- At the end of each monthly culture passage, separate or excise the developing clusters of axillary shoots (Plate 19) and transfer them to fresh medium of the same composition for further proliferation of axillary buds. Transfer up to 4 shoots per culture vessel.

Plate 19: Shoot proliferation in patchouli during multiplication stage

(e) *Rooting and Establishment of the Plantlets*

- Isolated shoots are placed on MS medium to encourage shoot elongation and/or spontaneous rooting.
- If after two monthly passages on MS medium, rooting is not observed, place them on MS medium supplemented with NAA to induce root formation.
- Well-rooted plantlets are carefully removed from the agar-solidified medium. The agar medium is gently rinsed off from the roots with water and the plantlets are placed in netpots with suitable media. Transfer the plantlets to a greenhouse. Care should be taken initially by careful watering and partial shading to get a high *ex vitro* survival percentage. Fig. 3 illustrates the micropropagation technique by proliferation of axillary buds.

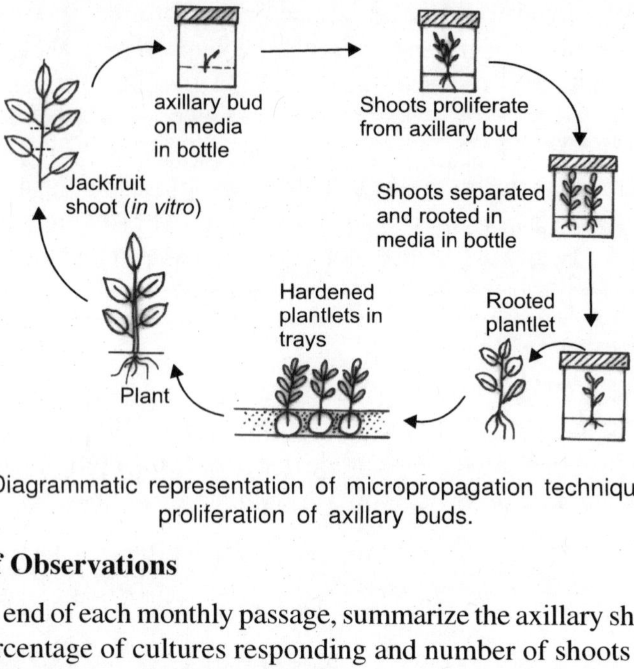

Fig. 3: Diagrammatic representation of micropropagation technique by the proliferation of axillary buds.

Schedule of Observations

1. At the end of each monthly passage, summarize the axillary shoot response by percentage of cultures responding and number of shoots developed.
2. Calculate the average multiplication rate on its optimum combination of explant and medium over the course of the experiment.

9.3. COMMON PROBLEMS ENCOUNTERED

Slow response: Some species in culture show a very slow response during the culture initiation phase, especially when the donor material is dormant. In such cases, a second monthly passage may be required before significant response is observed.

Some species exhibit a cyclical response, i.e., axillary shoots are initiated during one passage, but require a second passage for elongation and development before additional cycles of shoot proliferation can occur (e.g., Strawberry, Clove, Orchids).

Blackening or browning of tissues: Some kinds of plants, particularly tropical species, contain high concentrations of phenolic substances which are oxidized when cells are wounded or become senescent. The result is that isolated tissue becomes brown or black and fails to grow. Tissue blackening occurs through the action of copper containing oxidase enzymes such as Polyphenol oxidases and tyrosinase which are released or synthesized when presented with oxidative conditions when tissues are wounded. These enzymes and substrates (Tyrosine and hydroxy phenols such as chlorogenic acid) are normally retained within different compartments within cells and come together when cells are injured. Irreparable growth inhibition occurs when phenols are oxidized to highly active quinone compounds which then cyclize, polymerize and/or oxidize proteins to form melanic compounds. Blackening of tissues can often be prevented by one of the following approaches:

- Removing phenolic substances by leaching or dispersal
- By pretreatment of plant material by soaking in water (e.g., Seeds)
- By frequent transfer of the explants to fresh medium after a short time interval
- By initial culture in liquid medium so that the phenols diffuse away
- Adsorption with activated charcoal/polyvinylpyrolidone.
- Modifying redox potential (oxidation-reduction potential) by the addition of reducing agents (antioxidants) like ascorbic acid, citric acid, L-cysteine hydrochloride, dithiothreitol, glutathione and mercaptoethanol.
- Inhibiting the action of phenol oxidase enzymes by the addotion of chelating agents (NaFeEDTA, diethyldithiocarbamate and dimethyl-dithiocarbamate).
- Reducing phenolase activity and substrate availability by lowering the pH and by incubating the tissues in darkness. Some of the plants which exhibit extreme phenolic problems are mango, cashew, tamarind and to a considerable extent all perennial plants. Phenolic problem has been greatest impediment in mango tissue culture.

Vitrification: It is physiological condition wherein tissues appear water soaked or translucent. Vitrified tissues lack chlorophyll and become brittle, taking on a crystalline appearance. Other characteristics are rapid and abnormal growth, short internodes, thick curly leaves, loss of cuticle, excessive ethylene production, and death.

Vitrification is promoted by the type and concentration of agar and cytokinins in the medium. Weak agar and high cytokinins encourage vitrification. Chemically, the problem has been attributed to a deficiency of lignin, which is essential to the structure of cell walls and other tissues. Vitrification can be controlled by

- Use of minimum of cytokinins and higher amount of agar
- Maximizing the air exchange through the container closure
- The addition of phloroglucinol to the media
- Cold treatment has also been found to have some remedial effect

9.4. PROTOCOL FOR MICROPROPAGATION OF BRAHMI (*BACOPA MONNIERI* L. PENNELL) BY AXILLARY BUD PROLIFERATION (Dalia Mathews and Sathyanarayana, 2003)

Surface Sterilization and Explant Preparation

1. Obtain healthy and disease free shoots of Brahmi and keep them in running water for half an hour to remove all the dirt and debris present.
2. Soak the shoots in a surfactant solution (Tween-20) for 10 minutes.
3. Wash the shoots in distilled water, trim off the leaves and again wash several times in distilled water.
4. Transfer the shoots to 0.1% mercuric chloride for 2.5 minutes with frequent agitation in a sterile environment.
5. Wash the sterile shoots with several changes of sterile water to remove even traces of sterilent.

Culture Initiation and Maintenance

1. Cut the shoots into 1cm size segments each containing 1-2 nodes and place the prepared explants onto MS media supplemented with 2 mg/L BAP.
2. Four explants can be placed in each of the culture bottles.
3. Seal the culture bottles and incubate for three weeks.
4. At 3 weeks interval, excise the developing clusters of axillary shoots and transfer them to fresh media for further proliferation (Plate 20).

Plate 20: Multiple shoot formation from nodal segments of brahmi

Rooting and Establishment of the Plantlets

1. The plantlets root on the same media within 3-4 weeks and do not require transferring to auxin containing media.
2. Well-rooted plantlets are carefully removed from the agar-solidified media. The media is gently rinsed from the roots in running water and the plantlets are placed in netpots containing soilrite and sand in 1 : 1 ratio. The plantlets are immediately transferred to mist chamber.
3. The plantlets can be planted in the field after a week.

9.5. PROTOCOL FOR MICROPROPAGATION OF CURRY LEAF (*MURRAYA KOENIGII* SPRENG) FROM NODAL SEGMENTS (Prasad, 2000)

Collection and Preparation of Explants

- Collect young sprouts from healthy curry leaf plants. (September to October is the best season for collection of explants).
- Excise nodal segments measuring 2-2.5 cm in length and wash them thoroughly under running tap water for 15 minutes.
- Treat the explants with two drops of Tween-20 in 100 ml of distilled water for 20 minutes.
- Treat the explants with Bavistin 100 mg/L for 45 minutes and then rinse with sterile water.
- Disinfect the plants with 0.1% mercuric chloride for 10 minutes in a sterile environment.
- Rinse the explants in sterile distilled water for 2-3 times to ensure the complete removal of sterilents adhering to the explants.

Initiation of Cultures

- Inoculate the explants on MS media with 5mg/L of BAP for shoot proliferation (Plate 21).
- Incubate the cultures at a temperature of 25 ± 2°C and a photoperiodic regime of 16 hour light and 8 hour dark cycle with a light intensity of 1500 lux.

Plate 21: Establishment of nodal cultures of curry leaf

Rooting of Shoots

- Transfer well grown shoots to half strength MS medium with 3 mg/L NAA for *in vitro* rooting (Plate 22).

Plate 22: *In vitro* produced plantlets of curry leaf

9.6. PROTOCOL FOR MICROPROPAGATION OF JACKFRUIT (*ARTOCARPUS HETEROPHYLLUS* LAM.) (Sathyanarayana and Blake, 1993)

Preparation of explants

- Collect shoots from healthy plants. Remove all the leaves and cut into convenient sizes.
- Rinse thoroughly in forcefully running tap water for 30 minutes.
- Excise the nodal segments and sterilize them with 10% bleach (Sodium hypochlorite with 35% active chlorine) with 1 or 2 drops of wetting agent (Tween-20) for 10 minutes.
- If tip explants are utilized, they can be surface sterilized with 10% bleach with just a drop of the wetting agent for 5 minutes (as they are sensitive to prolonged contact with 10% bleach).
- Wash off the sterilent adhering to the plant material by giving two five-minute rinses with sterile double distilled water.

Culture Initiation

- Inoculate the sterile explants on half strength MS media with 2 mg/L BAP.
- Incubate the cultures at a temperature of 25 ± 20°C with 16h photoperiod.

In vitro Rooting of Shoots

- Transfer well grown shoots to half strength MS media with 1mg/L IBA for *in vitro* rooting (Plate 23).
- Incubate the cultures at a temperature of 25 ± 20°C with 16th photoperiod.

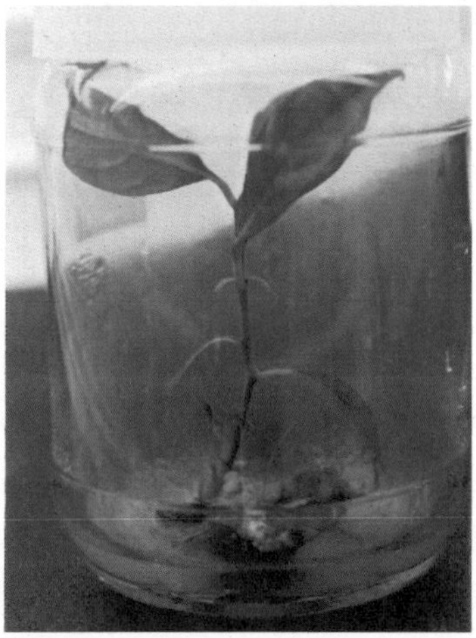

Plate 23: *In vitro* plantlet of jackfruit

Plate 24: Field planted tissue cultured jackfruit

9.7. PROTOCOL FOR *IN VITRO* PROPAGATION OF COLEUS (*COLEUS FORSKOHLII* BRIQ.) (Kathireshan, Farooqi and Sathyanarayana, 2002)

Explant Preparation

- Slender side shoots (1-1.5 cm long) from four-month-old plants are used as explants to initiate the cultures.
- Remove the expanded leaves of the shoots and wash the shoots with few drops of a surfactant (5% Teepol solution) in distilled water for 15 minutes at slow speed followed by thorough washing in running water for 30 minutes.
- Transfer the explants to 0.1% Bavistin solution for 45 minutes and then wash with sterile water.
- Surface sterilize the explants with 0.05% $HgCl_2$ for 9 minutes followed by 0.1% $HgCl_2$ for 4 minutes in a sterile environment.
- Thoroughly rinse with sterile water for 5 times to remove the traces of the adhering sterilent.

Culture Initiation and Maintenance

- Inoculate the explants on MS medium containing IAA (1.0 mg/L) and Kinetin (1.5 mg/L).

- Incubate the cultures at 25 ± 3°C with 16h photoperiod.
- Subculture every 4-6 weeks on media of the same composition (Plate 25).

Plate 25: Multiple shoot formation in coleus

In vitro Rooting of the Shoots

- Transfer individual shoots of 2-3 cm height to MS media with 1mg/L IAA for rooting of the shoots (Plate 26).

Acclimatization of the Rooted Plantlets

- Remove the well-rooted plantlets (Plate 26) after eight weeks of culture from the culture bottles and wash the roots free of agar. Plant them in plastic pots containing a mixture of sterile soil and sand (1:1) and transfer them to the green house.
- After one month, transfer the plants to the nursery and maintain under partial shade for a month after which they can be transplanted in the field.

Plate 26: *In vitro* rooted plantlets of coleus

9.8. PROTOCOL FOR DIRECT SHOOT REGENERATION FROM SHOOT TIPS OF GERBERA (*GERBERA JAMESONII*) (Vereena Lyngdoh and Sathyanarayana, 2002)

Explant Preparation

- Excise shoot tip explant of about 1cm size from *in vitro* derived plantlets. Trim off the leaves from the shoots.

Culture Initiation

- Inoculate the explants on MS media with 3mg/L BAP for culture initiation.
- Subculture at monthly intervals on the media with similar composition.
- Incubate the cultures in an air-conditioned room at a temperature of 27° ± 2°C under a photoperiod of 16 hours light and 8 hours dark cycle. Plate 27 shows multiple shoots obtained from shoot tip explants.

In vitro Rooting

- Place well-developed individual shoots on MS media with 7 mg/L IAA for rooting.

Acclimatization

- Remove well-rooted plantlets from the culture bottles and carefully wash off the agar adhering to the roots in tap water and then dip the shoots in Bavistin (0.1%) for 5 minutes.
- Plant them in netpots containing sand and soilrite (1 : 1) and transfer immediately to a mist chamber. After two weeks, transfer them to a greenhouse and after another two weeks, the plantlets can be shifted to the field.

Plate 27: Multiple Shoot formation in Gerbera

10

Micropropagation by Adventitious Shoot Regeneration

Plant regeneration from cultured tissues can also be achieved by culturing tissue sections lacking a preformed meristem (adventitious origin). Generally, adventitious regeneration events occur at unusual sites of a cultured tissue such as the internode, leaf blade, cotyledon or root elongation zone, where meristems do not naturally occur. However, adventitious shoot organogenesis shows the most variation as well as higher propagation rates.

10.1. OBJECTIVES

- To observe multiple shoot formation from leaf cultures of the given species.

10.2. EQUIPMENTS

- Laminar flow hood
- Forceps, scalpels
- Sterile beakers
- Sterile Petri dishes

10.3. MEDIA AND REAGENTS

- MS media supplemented with 2,4-D (0, 1, 2, 3, 4 and 5 mg/L) in combination with BAP 1 mg/L
- MS media supplemented with NAA 1mg/L
- Distilled water
- 70% alcohol
- Sodium hypochlorite/Mercuric chloride
- Agar (0.8%)
- Sterile water

10.4. PLANT MATERIALS

- Select healthy and vigorous shoots with healthy leaves showing no signs of disease or pest problems.
- Alternatively, *in vitro* shoot cultures can also be used.

10.5. PROCEDURE

10.5.1. Culture Initiation

- ➢ Excise healthy leaves from the shoots provided.
- ➢ Gently wash in warm water with a mild detergent, and rinse under tap water, followed by three rinses in distilled water.
- ➢ In the laminar flow chamber, surface sterilize the leaves using a suitable sterilent. (The concentration of the sterilent and period of treatment varies with the plant).
- ➢ Rinse the leaves in 5 changes of sterile distilled water.
- ➢ Place the leaves in a sterile Petri dish. Using sterile forceps and scalpel, cut each leaf into sections of 1sq cm, with and without vein in the explant.
- ➢ Place the leaf sections on each of the MS media.
- ➢ Seal each culture bottle and incubate for 4 weeks.

10.5.2. Shoot Multiplication

- At the end of the first monthly passage, identify the contamination free cultures for continued shoot multiplication.
- Transfer the culture material to a sterile Petri plate.
- With sterile forceps and scalpels, cut the mass of plantlets into smaller pieces and transfer to fresh MS media with similar composition.
- Seal the culture bottles and incubate them for 4 weeks.

10.5.3. Rooting of Shoots

- ➢ Excise and transfer the individual shoots to MS medium for shoot elongation and spontaneous rooting. If root initiation does not occur in two months time, transfer them to MS medium supplemented with NAA (1 mg/L).

10.5.4. Plantlet Establishment

- Gently remove well rooted plantlets from the culture vessel, keeping the roots intact.
- Transfer the plantlets with roots encased in agar media to a container of warm, but not hot, water and gently rinse the agar media off the roots.

- Plant them in small netpots with sterile soil mix. Make sure the media is moist with water, but is not sodden.
- Place the potted plantlets in misthouse followed by greenhouse and shadehouse so as to gradually harden them.

The micropropagation technique by direct adventitious shoot formation from leaf explants in brahmi is shown in Fig. 4.

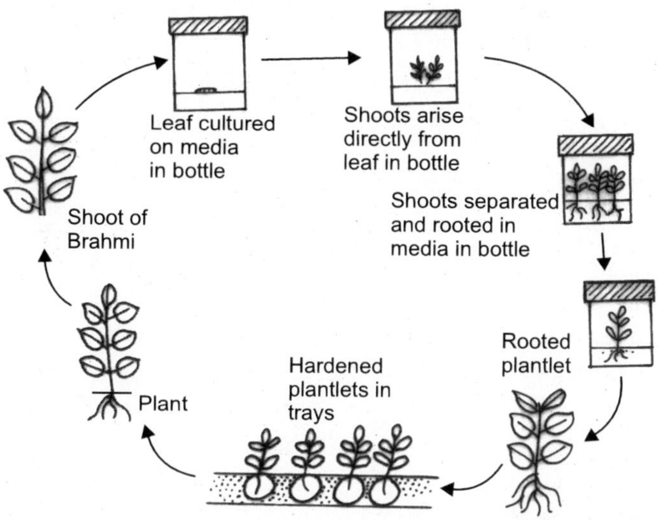

Fig. 4: Digrammatic representation of the micropropagation technique by direct adventitious shoot formation from leaf explant

10.6. SCHEDULE OF OBSERVATIONS

- Observe weekly for signs of contamination and calculate the percentage of contamination after a period of one month.
- Record the time of shoot bud emergence and the frequency of shoot bud formation.
- Record the time taken for root initiation and the frequency of root formation.

10.7. COMMON PROBLEMS ENCOUNTERED

10.7.1. Brown/Black Explants

Sometimes, the explants transferred onto the media may turn brown or black in colour within a coulpe of days without the potential to respond. This may be due to the death of the tissue. In such cases, reduce the strength of the detergent and/or decrease the time of treatment or concentration of the surface sterilant.

10.7.2. Lack of Shoot Multiplication

The lack of shoot multiplication after cutting the tissues into small pieces during subculturing may be due to the lack of a true shoot meristem in the subcultured pieces, in which case, the use of larger pieces should respond with further shoot multiplication and development. Lack of multiplication after subculture can also be due to the use of very small pieces of tissue, which literally have, no undamaged cells remaining that are capable of responding.

10.8. PROTOCOL FOR MICROPROPAGATION OF BRAHMI (*BACOPA MONNIERI* L. PENNELL) BY ADVENTITIOUS SHOOT FORMATION (Dalia Mathews and Sathyanarayana, 2003)

Surface Sterilization and Explant Preparation

1. Obtain healthy and disease free shoots with well-developed leaves of brahmi and keep them in running water for half an hour to remove all the dirt and debris present.
2. Soak the shoots in a surfactant solution (Tween-20) for 10 minutes.
3. Wash the shoots in distilled water several times.
4. Transfer the shoots to 0.1% mercuric chloride for 2.5 minutes in a sterile environment.
5. Wash the sterile shoots with several changes of sterile water to remove even traces of sterilent.

Culture Initiation and Maintenance

1. Remove leaves from the shoots and cut the internode segments, and place them onto MS basal media.
2. Four explants can be placed in each of the culture bottles.
3. Seal the culture bottles and incubate for four weeks. (Plates 28 and 29 show adventitious shoot formation from leaf and internodal explants, repsectively).

Rooting and Establishment of the Plantlets

1. The plantlets root on the same media within 3-4 weeks and do not require transferring to auxin containing media.
2. Well-rooted plantlets are carefully removed from the agar-solidified media. The media is gently rinsed from the roots in running water and the plantlets are placed in netpots containing soilrite and sand in 1 : 1 ratio. The plantlets are immediately transferred to mist chamber.
3. The plantlets can be planted in the field after a week.

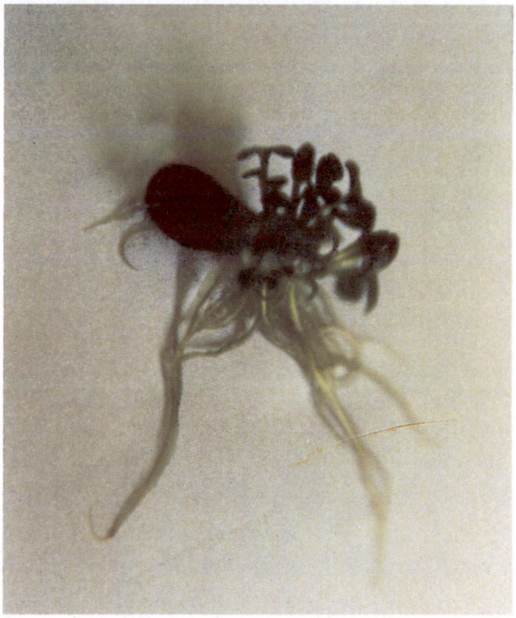

Plate 28: Adventitious shoot formation from leaf explants of brahmi

Plate 29: Adventitious shoot formation from internodal segment of brahmi

10.9. PROTOCOL FOR MICROPROPAGATION OF PATCHOULI (POGOSTEMON PATCHOULI) BY ADVENTITIOUS SHOOT FORMATION (Ramachandra, Farooqi and Sathyanarayana, 2003)

Surface Sterilization and Explant Preparation

- Collect leaves from mother plants maintained in the greenhouse and bring to the laboratory. Petioles from the leaves can also be used to induce adventitious plantlets.
- Wash in running water 4-5 times to remove all the adhering dirt and debris.
- Soak in a surfactant (Tween-20) for 10 minutes.
- Wash in distilled water and then immerse in a fungicidal solution (Bavistin 0.1%) for 30 minutes.
- Wash again in distilled water and surface sterilize with mercuric chloride (0.01%) for 4 minutes, in a sterile environment.
- Rinse with sterile water in as many changes as possible.

Culture Initiation and Maintenance

- Inoculate the explants on MS media with 1.0mg/L NAA + 0.1 mg/L BAP.
- Place 3-4 explants per culture bottle and incubate for 3-4 weeks.

Rooting and Establishment of the Plantlets

- Transfer well-developed shoots to half strength MS media for rooting
- Transplant well-rooted plantlets to sterilized media containing soilrite and sand in 1:1 ratio and harden the plantlets in amist house for a week followed by hardening in a greenhouse for another week.

10.10. PROTOCOL FOR DIRECT ADVENTITIOUS SHOOT FORMATION IN CARNATION (*DIANTHUS CARYOPHYLLUS* L.) (Al-Thobhani and Sathyanarayana, 1998)

Explant Preparation

- Collect shoots of 10 to 15 cm length from healthy, disease free plants. In the laboratory, excise nodal cuttings of 1-2 cm size.
- Wash the explants in running tap water. Immediately, soak them in a solution containing Benomyl (0.15%) and Citrimide (0.1%) along with two drops of a wetting agent, Tween-20. Agitate the explants on a shaker for 45 minutes.

- Disinfect the explants with 0.1% mercuric chloride solution for 10 minutes with continuous agitation inside the laminar flow chamber.
- Rinse the explants thoroughly with sterile distilled water 4 to 5 times.

Culture Initiation

- Cut the ends of the surface sterilized explants and inoculate onto MS media with 3 mg/L BAP + 0.2 mg/L NAA.
- Shift the culture bottles to the incubation chamber and incubate at a temperature of 25 ± 2°C and a photoperiodic regime of 16h light and 8h dark cycle (Plate 30).

Plate 30: Adventitious shoot formation in carnation

11

PLANT REGENERATION FROM CALLUS CULTURES

The first step towards *de novo* organogenesis is to establish callus or cell suspension cultures. Explant tissues generally show distinct planes of cell division, various specializations of cells and organization into specialized structures such as the vascular system. Callus formation from explant tissue involves the development of progressively more random planes of cell division, less frequent specialization of cells, and loss of organized structures.

Callus is a coherent but unorganized and amorphous tissue, formed by the vigorous division of plant cells. It is often induced in or upon parts of an intact plant by wounding or as a result of stress. Callus formed during *in vitro* culture has some similarities to tissue arising *in vivo* after injury to plants (so called wound callus). However, often there are differences in morphology, cellular structure, growth and metabolism between calli derived through tissue culture and natural wound callus. In culture, callus is initiated by placing small pieces of the whole plant onto a growth-supporting medium under sterile conditions. With the stimulus of endogenous growth substances or growth regulating chemicals added to the medium, cell metabolism is changed from being quiescent to metabolically active. The process of cell differentiation and specialization occurring in the intact plant is reversed and the explanted tissue is said to become dedifferentiated. During dedifferentiation, storage products typically found in resting cells tend to disappear and new meristems arise producing undifferentiated parenchymatous cells without any of the structural order that was characteristic of the organ or tissue from which they were derived. Lack of organization persists as the callus proliferates, although as growth proceeds, some kinds of specialized cells may again be formed at random.

The technique of callus culture was first developed in the late 1920s and early 1930s and was one of the primary methods of tissue culture for many years. Very few plant organs fail to respond to treatments designed to induce callus formation. Partly for this reason, callus cultures are very frequently used in a variety of biochemical, physiological and genetic experiments.

When subcultured regularly on agar media, callus cultures will exhibit sigmoidal pattern of growth during each passage. There are five phases of callus growth:

- A lag phase, where cells prepare to divide
- An exponential phase, where the rate of cell division is highest
- A linear phase, where cell division slows down, but the rate of cell expansion increases
- A deceleration phase, where the rates of cell division and elongation decrease, and
- A stationary phase, where the number and size of cells remain constant.

In the callus cultures, provision of a hormone signal induces the appropriate organ, either shoot or root. Hormone induction signals by providing cytokinins as well as auxins are required to form the complete plant having both shoot and root. Additional culture steps on hormone free medium may or may not be required following the use of the induction medium.

11.1. MEASUREMENT OF CALLUS GROWTH

Analysis of callus cultures is usually based on fresh weight or dry weight measurements. Measurement of fresh weight is a quick and easy method for assessing callus growth and does not involve destruction of the material. However, fresh weight measurements mainly reflect the water content of the tissue and may not accurately reflect the actual growth of the callus. Other measures that are used to assess callus growth include callus dry weight, cell number of the aggregate, mitotic index of the callus and respiration rate.

11.2. OBJECTIVES

- ➢ To initiate callus cultures from cotyledons
- ➢ To observe the induction of *de novo* shoot and root organogenesis

11.3. EQUIPMENTS

- Laminar flow hood
- Sterile forceps, scalpels
- Sterile Petri dishes

11.4. MEDIA AND REAGENTS

- MS media supplemented with 2,4-D (0, 1, 2, 3, 4 and 5 mg/L)

- MS media supplemented with 2,4-D (0.5mg/L) in combination with BAP (1 mg/L)
- MS media supplemented with NAA 1mg/L
- Distilled water
- 70% alcohol
- Sodium hypochlorite/Mercuric chloride
- Agar (0.8%)
- Sterile water

11.5. PLANT MATERIALS

- Healthy, well-developed seeds

11.6. PROCEDURE

11.6.1. Aseptic Germination of Seeds

- ➢ Wash the seeds by submerging in water with a few drops of liquid detergent in a flask and shake well for 2-3 minutes.
- ➢ Rinse in tap water to remove the detergent and then twice in distilled water.
- ➢ In a sterile laminar hood, decant the solution and treat the seeds with mercuric chloride solution (0.05%) for 4 minutes.
- ➢ Wash the seeds thoroughly in five changes of sterile distilled water.
- ➢ Germinate the seeds by transferring them to a moistened filter paper in a Petri dish.

11.6.2. Callus Initiation and Maintenance

- Collect the aseptically germinated seedlings when the cotyledons are fully expanded and the epicotyl is beginning to emerge.
- Excise the two cotyledons and culture them with their abaxial surface upside on the MS media supplemented with 2,4-D at different concentrations for callus initiation.
- Seal the culture vessels and incubate for 4 weeks.
- Once callusing occurs, excise small pieces of callus and subculture on fresh medium of the same composition every month to maintain a callus stock (Plate 31).

11.6.3. Induction of Shoots from Callus

- Place small pieces of the callus on MS media with combinations of 2, 4-D and BAP.
- Seal the culture vessels and incubate for 4 weeks.
- Observe the cultures weekly for signs of morphogenetic response. If the cultures show morphogenetic response, subculture onto fresh media of the same composition and incubate for another 4 weeks.

- Regenerated shoots can be rooted and hardened as per the protocol given in the previous experiment.

11.7. SCHEDULE OF OBSERVATIONS

- Determine the fresh weight of the callus produced from explants: place each callus piece inside a pre-weighed, sterile Petri dish to weigh the callus. Fresh weight of the callus is calculated after subtracting the weight of the Petri dish.
- Determine the dry weight of the calli: place the callus on an aluminium foil and dry the callus piece in an oven at 60°C for 48h and calculate the dry weight.

Plate 31: Friable callus from a plant species

- Note the type of calli produced (watery/friable/compact) and the colour of the calli.
- Record the percentage of calli showing morphogenetic response.

Fig. 5 Illustrates the micropropagation technique by indirect regeration.

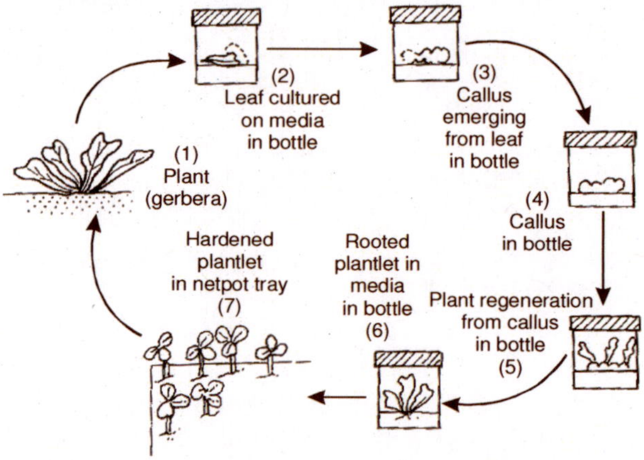

Fig. 5: Micropropagation by indirect regeneration

11.8. COMMON PROBLEMS ENCOUNTERED

If the callus formation is slow, its frequency can be enhanced by increasing the size of the explant or the surface area of the wound site. Many plants exhibit a strong tendency to develop callus in response to wounding.

11.9. PROTOCOL FOR MICROPROPAGATION OF BRAHMI (*BACOPA MONNIERI* L. PENNELL) BY INDIRECT ORGANOGENESIS (Dalia Mathews and Sathyanarayana, 2003)

11.9.1. Surface Sterilization and Explant Preparation

- Obtain healthy and disease free shoots of brahmi and keep them in running water for half an hour to remove all the dirt and debris present.
- Soak the shoots in a surfactant solution (Tween-20) for 10 minutes.
- Wash the shoots several times in distilled water.
- Transfer the shoots to 0.1% mercuric chloride for 2.5 minutes in a sterile environment.
- Wash the sterile shoots with several changes of sterile water to remove even traces of sterilent.

11.9.2. Culture Initiation and Maintenance

- Remove the internodes and leaves from the shoots and place the prepared explants onto MS media supplemented with 1mg/L 2,4-D + 0.1 mg/L BAP.
- Four explants can be placed in each of the culture bottles.
- Seal the culture bottles and incubate for four weeks. (Plate 32)

11.9.3. Shoot Regeneration from Calli

- Calli maintained on the same media regenerate shoots within a months time (Plate 33).

11.9.4. Rooting and Establishment of the Plantlets

- The plantlets root on the same media within 3-4 weeks and do not require transferring to auxin containing media.
- Well-rooted plantlets are carefully removed from the agar-solidified media. The media is gently rinsed from the roots in running water and the plantlets are placed in netpots containing soilrite and sand in 1 : 1 ratio. The plantlets are immediately transferred to mist chamber.
- The plantlets can be planted in the field after a week.

Plate 32: Callus from leaf explants of brahmi

Plate 33: Regeneration of shoots from calli

11.10. PROTOCOL FOR CALLUS INDUCTION AND SHOOT REGENERATION IN CARNATION (*DIANTHUS CARYOPHYLLUS* L.) (Al-Thobhani and Sathyanarayana, 1998)

Explant Preparation

- Collect leaves from healthy, disease free plants.
- Wash the explants in running tap water. Immediately, soak them in a solution containing Benomyl (0.15%) and Citrimide (0.1%) along with two drops of a wetting agent, Tween-20. Agitate the explants on a shaker for 45 minutes.
- Disinfect the explants with 0.1% mercuric chloride solution for 10 minutes with continuous agitation inside the laminar flow chamber.
- Rinse the explants thoroughly with sterile distilled water 4 to 5 times.

Callus Induction

- Cut the surface sterilized explants into pieces of about 1 cm length and inoculate onto MS media with 1 mg/L 2,4-D.
- Shift the culture bottles to the incubation chamber and incubate at a temperature of 25 ± 2°C and a photoperiodic regime of 16h light and 8h dark cycle (Plate 34).

Plate 34: Callus induction from leaves of carnation

Shoot and Root Regeneration from Callus

- Subculture the callus onto MS media containing 10 mg/L of BAP for shoot and root regeneration.
- Incubate at a temperature of 25 ± 2°C and a photoperiodic regime of 16h light and 8h dark cycle.

11.11. PROTOCOL FOR CALLUS INDUCTION IN CHILLI (*CAPSICUM ANNUUM L. VAR ACUMINATUM*) (Mathew and Sathyanarayana, 2000)

Plant Material Used

- Seeds and leaf bases serve as the explants.
- Collect leaves from 30-35 day old *in vitro* raised seedlings. Cut the leaf into two pieces. The basal part of the leaf (0.5 cm long) along with leaf petiole measuring about 0.5 cm is used as the explant.

Surface Sterilization of the Leaf Bases

- Wash the leaves under running tap water for 15 minutes.
- Place the explants in distilled water with two drops of Tween-20 (a surfactant) for 20 minutes with vigorous shaking followed by washing in sterile water.
- Treat the explants with 0.05% mercuric chloride for 2 minutes followed by rinsing in sterile water five times in an aseptic environment.

Surface Sterilization of Seeds

- Wash the seeds thoroughly in running water and treat with Tween-20 for 15 minutes.
- Soak the seeds in 1% Bavistin for 45 minutes followed by washing in sterile water.
- Surface sterilize the seeds with 0.05% mercuric chloride solution for 5 minutes in the laminar flow chamber.
- Rinse the seeds in sterile water several times to remove traces of the sterilent.

In vitro Germination of Seeds

- Inoculate the sterile seeds on basal Murashige and Skoog's semi-solid agar medium supplemented with 3% sugar.
- Keep the bottles in the growth room under dark conditions.
- Transfer the germinated seeds with hypocotyl and cotyledons to light conditions.

Callus Induction

- Inoculate the leaf bases and hypocotyl region of the seedlings on MS media with 3 mg/L 2,4-D and 0.5 mg/L BAP for callus induction. (Plate 35 and 36)
- Incubate at a temperature of 25 ± 2°C and a photoperiodic regime of 16h light and 8h dark cycle.

Plate 35: Callus induced from leaf bases of chilli

Plate 36: Callus from hypocotyls region of chilli

11.12. PROTOCOL FOR CALLUS INDUCTION AND SHOOT REGENERATION IN *ANTHURIUM ANDREANUM* LIND. (Nirmala and Singh, 1996)

Explants Used

Whole leaf, spadix, petiole and inflorescence stalk

Explant Preparation

- Collect explants from healthy and disease free explants.

- Disinfect them in 70% ethanol for a few seconds followed by washing in solution of Labolene (0.1ml/L) + Tween-20 (0.1ml/L).
- Immerse in freshly prepared chlorine water, diluted with sterile water in 1 : 1 ratio for 30 minutes in a sterile environment.
- Rinse with sterile water three times to remove the traces of chlorine.
- Cut the leaves into sections of 1cm^2 and the inflorescence stalk, spadix and petiole into 0.5-1cm segments.

Culture Initiation

- Inoculate the leaf sections on modified MS media (with Ammonium nitrate concentration reduced to half) with 1.0 mg/L BA + 0.2 mg/L 2,4-D. (Leaf sections with their adaxial surface placed on the medium gives better callusing).
- Inoculate the petiole and inflorescence stalk segments also on modified MS media with 1.0 mg/L BA + 0.1 mg/L 2,4-D.
- Incubate the cultures in dark for callus induction (Plate 37).

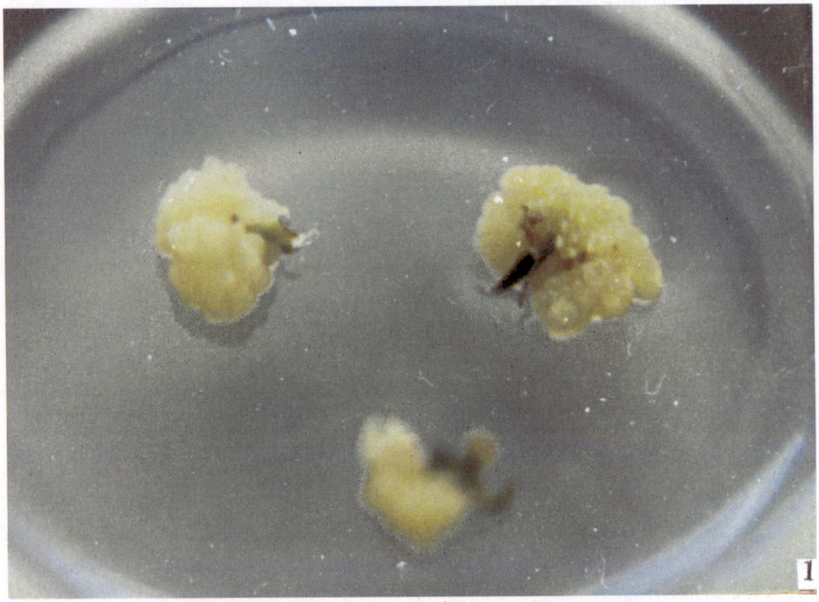

Plate 37: Callus induced from Anthurium leaves

Shoot and Root Regeneration

- Inoculate callus pieces on MS media with 0.5mg/L kinetin for shoot and root regeneration (Plate 38).

Plate 38: *In vitro* derived plantlets of Anthurium

11.13. PROTOCOL FOR *IN VITRO* PROPAGATION OF COLEUS (*COLEUS FORSKOHLII* BRIQ.) BY INDIRECT ORGANOGENESIS (Kathireshan, Farooqi and Sathyanarayana, 2002)

Explant Preparation

- Healthy, disease free leaves from four-month-old plants are used as explants to initiate the cultures.
- Remove the fully expanded leaves from the shoots and wash them with few drops of a surfactant (5% Teepol solution) in distilled water for 15 minutes at slow speed followed by thorough washing in running water for 30 minutes.
- Transfer the explants to 0.1% Bavistin solution for 45 minutes and then wash with sterile water.
- Surface sterilize the explants with 0.05 % $HgCl_2$ for 9 minutes followed by 0.1% $HgCl_2$ for 4 minutes in a sterile environment.
- Thoroughly rinse with sterile water for 5 times to remove the traces of the adhering sterilent.

Culture Initiation and Maintenance

- Cut the leaves into small pieces of $1 cm^2$ size and inoculate on MS medium containing BAP (1.5 mg/L) and 2,4-D(1.0 mg/L).
- Incubate the cultures at 25 ± 3°C with 16h photoperiod.
- The calli will regenerate shoots on the same media (Plate 39).

Plate 39: Regeneration from callus of coleus

In vitro Rooting of the Shoots

- Transfer individual shoots of 2-3 cm height to MS media with 1mg/L IAA for rooting of the shoots.

Acclimatization of the Rooted Plantlets

- Remove the well-rooted plantlets after eight weeks of culture from the culture bottles and wash the roots free of agar. Plant them in plastic pots containing a mixture of sterile soil and sand (1:1) and transfer them to the greenhouse.
- After one month, transfer the plants to the nursery and maintain under partial shade for a month after which, they can be transplanted in the field.

11.14. PROTOCOL FOR CALLUS INDUCTION AND SHOOT REGENERATION IN GERBERA (*GERBERA JAMESONII*) (Vereena Lyngdoh and Sathyanarayana, 2002)

Explant Preparation

- Excise healthy leaves and wash in tap water for 30 minutes to remove all the soil adhering to it followed by washing in distilled water containing 2-3 drops of Tween-20. Rinse in distilled water twice.
- Treat the explants with Bavistin (0.01%), a fungicide, for 10 minutes, followed by treatment with streptomycin (0.01%). Rinse the leaves twice in distilled water.
- Surface sterilize the explants in 2% sodium hypochlorite solution (i.e., 50ml of sodium hypochlorite in 50ml distilled water) in the laminar flow for 7-9 minutes followed by washing in sterile water three times.

Culture Initiation

- Trim the leaves to 2-4 mm^2 size and inoculate on MS media with 7 mg/L BAP and 0.5 mg/L IBA for callus initiation.
- Incubate the cultures in an air-conditioned room at a temperature of 27° ± 2°C under dark conditions (Plate 40).
- Subculture the calli on to MS media with similar composition for shoot and root regeneration.

In vitro Rooting

- Place well-developed individual shoots on MS media with 7 mg/L IAA for rooting.

Plate 40: Callus induced from Gerbera leaves

Acclimatization

- Remove well-rooted plantlets from the culture bottles and carefully wash off the agar adhering to the roots in tap water and then dip the shoots in Bavistin (0.1%) for 5 minutes.
- Plant them in netpots containing sand and soilrite (1 : 1) and transfer immediately to a mist chamber. After two weeks, transfer them to a greenhouse (Plate 41) and after another two weeks, the plantlets can be shifted to the field.

Plate 41: Gerbera plantlets in the net pot at greenhouse

12

PLANT REGENERATION BY SOMATIC EMBRYOGENESIS

In plant tissue culture technique, the developmental process of embryo formation from the somatic plant tissue is known as somatic embryogenesis. In somatic embryogenesis, somatic cells develop by division to form complete embryos analogous to zygotic embryos. The callus tissue is modified into organized small bipolar structure which later on bears cotyledons resembling zygotic embryo. The bipolar structure of the somatic embryo contains both shoot and root meristems. As the embryo develop, they progress through the distinct structural steps of the proembryo, globular, heart, torpedo and mature stages. The various stages in somatic embryo development are usually described as follows:

Proembryo : Small clusters of meristematic cells from which somatic embryos will arise

Globular stage : Larger groups of cells not yet having a definite embryoid shape (Plate 42)

Heart stage : A characteristic three-lobed stage where cotyledonary initials are separated from the root pole

Torpedo stage : An elongated form of the heart-shaped embryo

Mature stage : Discernible small 'seedling' with primary root and shoot

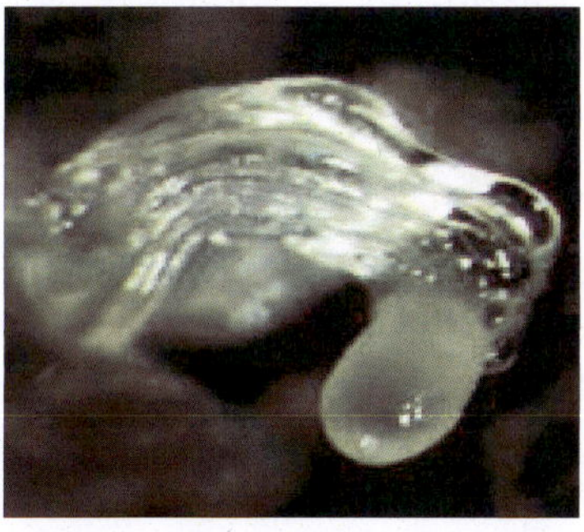

Plate 42: Globular stage showing connective

The process of somatic embryogenesis may be exemplified in two ways: direct and indirect. Direct somatic embryogenesis is the formation of somatic embryos or embryogenic tissue directly from the explant without the formation of an intermediate callus phase (Raghavan, 1986). To induce direct embryogenesis, the explant most often used is the immature zygotic embryo (Merkle *et al.*, 1990). This tissue is already embryogenic in nature and apparently requires less nurturing than other somatic tissues to induce a somatic embryogenic response. The process of indirect somatic embryogenesis was explained by several workers particularly Reinert in 1959. Here, the excised somatic cells are placed in the culture medium aseptically. The cells undergo callus proliferation and the embryo like structures are produced from induced embryogenic cells.

Somatic embryogenesis in many plant species is initiated in the same manner as the production of callus. Induction of somatic embryogenesis in most species requires a high concentration of auxins, usually 2,4-D in the culture medium. Cytokinin usually is not required for the induction of somatic embryogenesis, but certain monocot species do have a specific requirement for cytokinin.

The high concentration of auxin used for induction, however, is usually inhibitory to the development of the somatic embryos into advanced stages. Hence, a hormone free medium often is used for the development of globular-stage somatic embryos into plantlets. Sometimes, low concentration of hormones in the medium can be beneficial or even required, depending on the species, to promote normal development of the embryos.

***Pseudobulbils*:** At times, highly embryogenic calluses are formed consisting entirely of globular cell masses, which bud off new ones. Such globular bodies tend to reach a larger size under conditions that are not conducive to embryogenesis. These structures are often referred to as pseudobulbils.

12.1. OBJECTIVES

- To initiate and maintain callus cultures of the plant given
- To observe the induction and development of somatic embryos from the cultured cells

12.2. EQUIPMENTS

- Laminar flow hood
- Sterile forceps and scalpel
- Sterile Petri dishes

12.3. MEDIA AND REAGENTS

- MS basal media

- MS media supplemented with 2,4-D (2 mg/L)
- MS media supplemented with 2,4-D (1 mg/L) + BAP (0.5 mg/L)
- Distilled water
- 70% alcohol
- Sodium hypochlorite/Mercuric chloride
- Agar (0.8%)
- Sterile water

12.4. PLANT MATERIAL

- Aseptically germinated seedlings to initiate callus culture

12.5. PROCEDURE

(*a*) *Callus Initiation and Maintenance*

- Calli can be initiated from cotyledons as well as the hypocotyls region, as per the protocol given in the previous exercise (MS media with 2 mg/L, 2, 4-D is used for callus induction).
- Small pieces of calli can be maintained by subculturing on fresh medium of the same composition.

(*b*) *Induction of Somatic Embryos*

- Place small pieces of the calli on the MS media with 2, 4-D (1 mg/L) + BAP (0.5 mg/L) and on MS to basal media.
- Seal the culture vessels and incubate for 4 weeks.
- Observe the cultures weekly for signs of somatic embryogenensis.
- Regenerated plantlets can be isolated for further development. The rooted plantlets can be transferred to soil for *ex vitro* establishment.

Schedule of Observations

- At the end of each monthly passage for callus, observe for embryogenic response. Calculate the percentage of cultures expressing embryogenesis.
- Calculate the number of somatic embryos per gram of fresh weight of callus and categorize the embryos by stage of development, i.e., globular, heart, torpedo, etc.

12.6. COMMON PROBLEMS ENCOUNTERED

12.6.1. Abnormal Somatic Embryos

A common problem in some plant species is the occurrence of abnormal somatic embryos. The cotyledons may be twisted, one cotyledon may be more developed than the other, the shoot meristem may not be well formed or may be absent, etc. These systems often require additional cultural manipulation to promote more normal appearance and development of somatic embryos such as the use of low to moderate concentrations of auxins, cytokinin, and/or gibberellin in the expression medium, use of abscisic acid or high osmaticum treatments or other physical treatments such as desiccation.

12.7. SYNTHETIC SEEDS/ARTIFICIAL SEEDS

A synthetic seed contains an embryo produced by somatic embryogenesis, enclosed within an artificial medium that supplies nutrients and is encased in an artificial covering.

12.7.1. Advantages of Artificial Seeds

- Helps in clonal propagation to replace traditional seed propagation. A replacement for hand pollinated hybrid plant
- Carriers for beneficial microorganisms, pesticides and growth regulators
- Not season dependent
- Role of different embryonic constituents can be studied

12.7.2. Protocol for Encapsulation of Seeds

Common chemicals used for encapsulation are sodium alginate with gelatin (2.0% w/v), Gelrite (0.25% w/v), calcium chloride (500 mM), etc. There are two main methods of encapsulation of seeds:

12.7.2.1. Gel Complexation via Dropping Procedure

1. After the induction of somatic embryogenesis, the embryos are isolated by suspension culture.
2. The isolated embryoids are kept in a beaker with 0.5-5.0% sodium alginate solution and the beaker is incubated.
3. Each embryoid is then dropped into 30-100mM calcium nitrate solution for 30 minutes so that the embryoids are gelled completely.
4. After encapsulation, the embryoids are placed aseptically in a culture flask and the culture is kept in dark for a few weeks.
5. The embryoids can be then transferred to the greenhouse.

12.7.2.2. Alternate Method

In the alternate method, the isolated embryoids are mixed with gelrite in a beaker. It is then placed in a well of a microtiter plate. As the temperature reduces, the encapsulation of the embryoids occurs, as solidification of gelrite is temperature dependent.

12.8. PROTOCOL FOR SOMATIC EMBRYOGENESIS AND PLANTLET REGENERATION FROM NODE CALLUS CULTURES OF *GLORIOSA SUPERBA* L. (Kalpana Sachdev, Prakash, Prasad and Sathyanarayana, 2002)

Explant Preparation

- Select node explants from healthy *in vivo* plants.
- Subject the explants to running water treatment for 30 minutes followed by washing in double distilled water (DDW).
- Rinse with 70% (v/v) ethanol and then agitate in a solution containing 100 ppm Bavistin + 500 ppm cetrimide for one hour.
- Decant off the solution and wash the explants several times with DDW.
- Surface sterilize the explants with 0.1% mercuric chloride solution in the laminar airflow chamber. Rinse thoroughly with sterile water.

Callus induction

- Culture the explants on MS medium amended with 4.0 mg/L 2, 4-D.
- Incubate at a temperature of 25 ± 2°C.

Induction of Somatic Embryos

- Subculture the induced calli obtained from nodal explants in MS media with 1.0 mg/L NAA + 6.0 mg/L BAP.
- Further subculture this calli on MS media with 4.0 mg/L BAP for further development of the somatic embryos and plantlet regeneration. (Plate 43)

Plate 43: Somatic Embryogenesis in *Gloriosa superba L.*
A. *Embryogenic callus* B. *Stages of somatic embryogenesis* C. *Plantlets from embryos.*

Time scale (Months)	Culture Stage	Number of propagules in the culture vessel	Remarks
2	Stage I	50 stem nodes cultured in 4 mg/L 2, 4-D 80% success — 60 days Callus from 40 nodes divided into 120 clumps of 1 gm each 110 best clumps were used 45 days	Disinfection and inoculation of node segments in test tubes Establishment of cultures in test tubes
1.5	Stage II	Somatic embryos produced as clumps in 6 mg/L BAP + 1 mg/L NAA (These were divided into 110 clumps, 1gm each.) 60 days	Callus was transformed to variously shaped somatic embryos in test tubes
2	Stage III	35 plantlets per clump × 110 clumps = 3850 shoots in 4 mg/L BAP 1st Cycle — 30 days	The somatic embryos germinated only in BAP treatments giving rise to shoots in bottles
1	Stage IV	3850 shoots × 13 = 49050 rooted shoots in 2.5 mg/L BAP + 0.25 mg/L IBA 30 days	2nd Cycle Rooted multiple shoots → 49050 × 13 = 637650 shoots in 2.5 mg/L BAP + 0.25 mg/L IBA
1	Stage V	49050 rooted plantlets in 10 mg/L Kinetin 60 days	Tuber induction — 637650 rooted plantlets in 10 mg/L Kinetin
2	Stage VI	49050 hardened plants	637650 hardened plantlets

Fig. 6: Flow diagram of a self-sustaining micropropagation system for *Gloriosa superba* L. reproduced after Sachdev, *et al.*, 2002

Multiple Shoot Production

- Transfer the regenerated clumps of plantlets to MS medium with 2.5 mg/L BAP and 0.25 mg/L IBA for multiple shoot production.

In vitro Tuber Production

- To induce tubers, divide the regenerated shoots into small clumps and transfer to MS media with 10.0 mg/L kinetin.

Acclimatization of the Plantlets

- Carefully lift the regenerated shoots with well-developed roots and tubers from the culture bottles and wean out into plastic pots containing equal quantities of farmyard manure, peat and perlite.
- Maintain the plantlets at 25 ± 2°C and 90% relative humidity.
  Fig. 6 gives the flow diagram of a self-sustaining micropropagation system for *Gloriosa superpa.*

12.9. PROTOCOL FOR DIRECT SOMATIC EMBRYOGENESIS IN SWEET TYPE CARAMBOLA (*AVERRHOA CARAMBOLA* L.) (Sathyanarayana, Prasantha and Sondur, 2002)

Explants Used

- Excise root explants from 5-6 weeks old *in vitro* grown seedlings of sweet type clones.

Induction of Somatic Embryos

- Inoculate the explants on half strength MS media supplemented with 2 mg/L BAP.
- Incubate the culture in a growth room provided with 16h of light conditions. (Plate 44 and 45)

Development of Plantlets

- Transfer the pro-embryos to MS media with IBA (1 mg/L).
- First roots arise, followed by shoots to get the complete plantlets.

Plate 44: Stages of somatic embryogenesis in sweet carambola

Plate 45: Somatic embryogenesis in sweet carambola, and note shoot organogenesis from embryos

13

Meristem Culture

Meristem culture is a method in which shoot apices containing the apical dome with a few primodial leaves are grown *in vitro*. This method is also known as apical tip culture.

Excision of the meristem and regeneration of plantlets on hormone amended media have permitted the propagation of plants free from systemically invasive viruses and from other disease causing pathogens. The application of meristem culture to obtain virus-free plants was based on the initial observation that apical tip of shoots and roots are generally free of viruses and other infection, mitosis in the meristem cells compete with the virus multiplication, high auxin level in the meristem may inhibit virus multiplication and tissue injury caused by meristem excision may result in degradation of the virus particles.

The meristem, which is about 0.1 mm in diameter and about 0.25 mm long, is composed of actively dividing and undifferentiated cells at the apices of shoots and roots. This tissue has no vascular system and is therefore, less likely to be infected with systemic viruses. The size of the initial explant is, however, an important factor. The smaller the size of the explant, greater is the chances of virus elimination, but lesser the chances of its survival.

To ensure complete elimination of viruses, it is often necessary that the plants are heat treated prior to meristem culture. The parent plants or cuttings are grown in a controlled temperature cabinet at 30°–40°C for 6-12 weeks till the new shoots develop. Following this treatment, the explants are cultured. When added to the culture medium or tissue treatment prior to culture, certain chemicals like virazole has also proved effective in freeing the tissue of viruses.

In spite of the treatments, virus indexing is usually done to ensure the establishment of a virus-free nuclear-stock and to detect any infection from time to time.

Morel and Martin undertook the pioneering work on meristem culture in

1952. They first demonstrated the concept of regenerating virus-free plantlets from meristem tips of virus-infected Dahlias. Since this pioneering work, meristem culture has been extensively used for the production of pathogen-free plants in asparagus, banana, cassava, cauliflower, gooseberry, pea, potato, raspberry, rhubarb, strawberry, sweet potato and yams.

13.1. OBJECTIVES

- To establish plantlets *in vitro* by nodal culture to provide the explant source
- To excise meristems from *in vitro* plantlets and to regenerate plantlets from meristems

13.2. EQUIPMENTS

- Laminar airflow hood
- Stereo microscope with illuminator for meristem excision
- Sterile culture tubes, Petri dishes
- Sterile forceps, scalpels

13.3. MEDIA AND REAGENTS

- MS basal media
- MS media supplemented with Kinetin (0.01 mg/L) + GA_3 (0.1 mg/L) (meristem medium)
- Distilled water
- 70% alcohol
- Sodium hypochlorite/Mercuric chloride
- Agar (0.8%)
- Sterile Water

13.4. PROCEDURE

(a) Establishment of *in vitro* Plants

➢ Establish plantlets by nodal cuttings on MS basal media and grow up to about six node stage.

(b) Meristem Excision

➢ Make nodal cuttings from the third and fourth nodes of respective plants

➢ Place one cutting under the dissecting microscope at 10-20× magnification and using scalpel and forceps, peel away the protective leaves on the bud

- ➢ Excise the meristematic dome plus the first set of leaf primordia with a scalpel
- ➢ Place the excised meristem on the meristem medium
- ➢ If the growth is not evident in 2-4 weeks, transfer the meristem to fresh medium
- ➢ Grow the plantlets from meristems to the six node stage
- ➢ Rooting and plantlet establishment are performed as in other experiments.

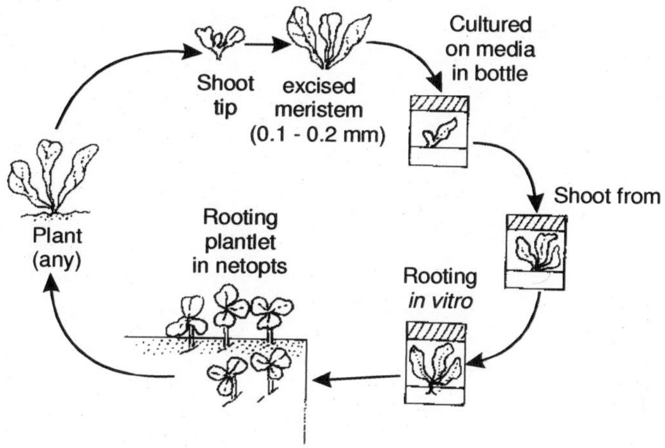

Fig. 7: Diagrammatic representaion of meristem culture

Schedule of Observations

- Record the survival percentage of the plantlets
- If possible, test the plantlets for virus by ELISA

13.5. PROBLEMS AND THEIR SOLUTIONS

13.5.1. Poor Growth of Meristems

Fresh meristem medium is important. If the surface of the medium dries out, the meristem will not grow. At times, the extremely small pieces of tissue may desiccate quickly in the laminar flow hood. Hence, it is critical to place them on the medium as soon as possible.

It makes sense to choose meristem explants from plant species in which the size of the meristem obtained is generally bigger, for example, banana. To be successful in meristem culture, students may opt to work initially on isolating as minute a shoot tip as possible greater experience or practice of isolating meristems is a must to be successful in meristem culture. A microscope of high precision and enlargement such as stereomicroscopes would be worthwhile possession in the laboratory.

14

In Vitro Haploid Production

Haploids are the plants with gametic chromosome constitution in the sporophyte. As early as 1922, Blackslee and his associates, while working with *Datura stramonium* reported the natural occurrence of haploids. Since then the number of species in which haploids have been reported to occur spontaneously by parthenogenesis or have been induced experimentally by various techniques has increased manifold.

Haploids may be grouped into two broad categories.

(*a*) Monoploids (monohaploids), which possess half the number of chromosomes from a diploid species.

(*b*) Poly haploids, which possess half the number of chromosomes from a polyploid species.

Recently, the interest in haploids has been increased substantially on account of its versatile uses in crop improvement. The usefulness of haploids is based on the assumption that homozygous inbred lines can be rapidly and readily achieved. Moreover, it is easier to detect mutations and variations in haploids because of the presence of a single set of chromosome. *In vitro* production of haploids can solve some problems in genetic studies since gene actions are readily manifested due to a single allelic dose present in chromosomes of an entire genome. This also makes them good material for the induction of mutations and various gene manipulations. Through rapid achievement of homozygous traits in double haploids, pollen derived haploid plants have been used in breeding and improvement of crop species.

14.1. USES AND APPLICATIONS OF ANTHER CULTURE

Anther/pollen cultures are useful in obtaining haploid plants. Haploids are required for two main reasons:

1. Isogenic/homozygous diploids can be obtained by chromosome doubling
2. The presence of one set of chromosomes facilitate the isolation of mutants

Haploids are of considerable value in obtaining homozygous diploids in short period of time. Selection based on haploids may be 6-8 times as efficient as that based on zygotic diploids.

A doubled haploid strain in self-pollinated crops

1. May be directly released as a variety if it performs well in yield trials.
2. It may be ordered to hybridization programme as a parent if it has some useful characters and hybrids thus obtained could be subjected to hybrid sorting.

A doubled haploid line in cross-pollinated crops can be used to make inbred line (if it is not there or difficult to produce, e.g., *Asparagus, Brassica*) or another inbred line in crops where already inbreds are developed (e.g., Maize). But these inbreds are not directly released as a cultivar but used for hybrid production.

Haploid plants are very important in the production of inbreds especially in the case of self-incompatible crops. Haploids from self-incompatible plants have the following uses:

- Use of completely homozygous diploid plants of self-incompatible plants as parental material for hybrid seed production.
- For the production of haploid plants from interspecific hybrids between self-compatible and self-incompatible species.

Haploid plants are used for developing new cultivar with improvements in yield and/or disease resistance, maturity duration, adaptability, etc.

***Hybrid sorting*:** Anthers from F1 hybrids from selected crosses are cultured to produce haploids. The chromosome number of these haploids is then doubled to get homozygous doubled haploid plants. This is referred to as hybrid sorting. Since the heterozygous gene combination present in hybrids is quickly resolved into homozygous combinations, hybrid sorting not only saves valuable time in the homozygous lines, but also improves the effectiveness of selection, which is now based on genotype only.

Haploid plants are also useful material for the study of somatic cell genetics, especially for mutation and cell modification. Isolated pollen and cells can be handled like microbes and treated with various mutagenic chemicals and irradiations. The advantage of haploid cells is that they can be easily diploidized to form homozygous diploids.

Haploid plants are useful in screening the mutants. Majority of the mutations are recessive and therefore are not expressed in diploid cells in the presence of an unmutated dominant gene. But this could be easily detected by haploid production and their diploidization.

In plants like *Freesia*, which are normally propagated vegetatively by corms, it takes 8-10 years to produce a clone large enough for commercial purpose. In these, anther culture has an advantage, specific desirable gene combinations may be linked in a haploid form and then converted to the fertile homozygous diploid or polyploid form in a relatively short time.

In addition, haploids could be used in

- Recurrent selection
- Fixing quantitative characters so that they can be analyzed and selected
- Identification of superior crosses.

Haploid protoplasts can also be used as a means of propagation of rare haploid plants.

Gametoclonal variation: Haploid plants regenerated from callus show genetic variation, which can be manipulated for crop improvement. This variation is termed as gametoclonal variation and are more desirable than somaclonal variation because:

- The mutant characteristic is expressed in the plants regenerated from the cells *in vitro*.
- Cells having detrimental mutations may be expected to regenerate much less frequently in the case of haploid cells than in diploid cells.

Haploid plants obtained from the pollen grains are being used for the improvement of tobacco where, disease resistant and promising lines have been developed through this technique.

Haploids are of considerable value in ornamentals wherein, plants with smaller flowers and extended flowering period is obtained through haploids.

14.2. *IN VITRO* METHODS FOR THE INDUCTION OF HAPLOIDS

There are mainly two methods of haploid plant production. These include:

1. Androgenic methods (anther and pollen culture)
2. Gynogenic methods (ovary and ovule culture)

14.3. ANDROGENIC HAPLOID PRODUCTION

There are three methods of androgenic haploid production.

(*a*) Anther culture

(*b*) Isolated pollen culture

(*c*) Haploid protoplasts

14.4. ANTHER CULTURE

Anther culture is the culture of complete anthers containing immature pollen microspores. Fig. 8 illustrates the technique of anther culture. The objective is usually to obtain haploid plants by the formation of somatic embryos directly from the pollen, or sometimes by organogenesis *via* callus. This is a more popular *in vitro* method of haploid production than the gynogenic pathway. The first report

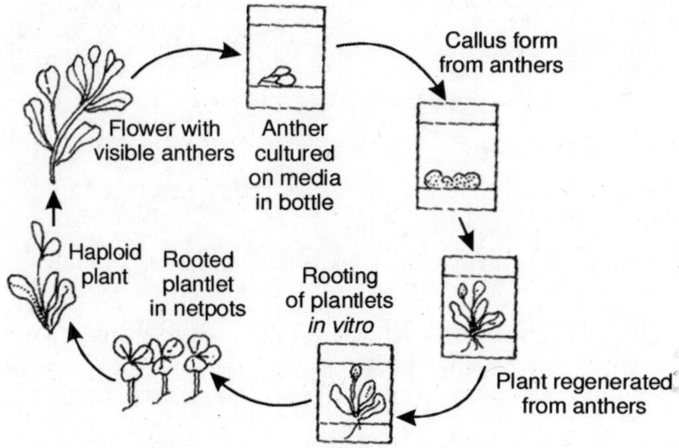

Fig. 8: Diagrammatic representation of anther culture

of haploid production was by Indian scientists Guha and Maheshwari in 1964 from isolated anthers of *Datura innoxia*. Bourgin and Nitsch in 1967 obtained haploids from anthers of *Nicotiana*. During the last decade, anther culture of rice, wheat, maize, *Brassica*, pepper and other crop species have been reported.

14.5. ONTOGENY OF ANDROGENESIS

Pollen grains of the cultured anthers show a remarkable cytological change during the first 6-12 days called as the inductive period. In tobacco, the gametophytic cytoplasm of the binucleate pollen grains are degraded, ribosomes are eliminated and only a few mitochondria and plastids remain. New ribosomes are synthesized following the first sporophytic division of the vegetative cell.

The pollen grains of many crop species (e.g., tobacco, barley, wheat, etc.) exhibit pollen dimorphism. In such cases, most of the pollen grains are bigger and they stain deeply with acetocarmine and contain plenty of starch. But only a very small proportion ca. 0.07% of the pollen grains are smaller and stain faintly with acetocarmine, these are called S grains. It is the S grains that respond during anther culture, the frequency of responding pollen grains can be enhanced over that or S grains by certain pretreatments e.g., chilling.

The early divisions in responding pollen grains may occur in one of the four following ways:

(*i*) The uninucleate pollen grain may divide symmetrically to yield two daughter cells both of which undergo further divisions.
Example: *Datura innoxia*

(*ii*) The uninucleate pollen divides unequally (as it does in nature). The generative nucleus degenerates immediately or after going through one or two divisions. The callus/embryo originates due to successive division of the vegetative cells

Example: *Nicotiana tabacum*, *Datura metal*, barley, wheat, triticale, chillies, etc.

(*iii*) The pollen embryos originate from the generative cell alone. The vegetative cell either does not divide or divides only to a limited extent forming a suspensor like structure.

Example: *Hyoscyamus niger*

(*iv*) The uninucleate pollen grains divide unequally producing generative and vegetative cells. But, both these cells divide repeatedly to contribute to the developing embryo.

Example: *Datura innoxia*

The responsive pollen grains may become multicellular and ultimately bursts open to release the cell mass. This cell mass may either assume the shape of a globular embryo and undergo the developmental stages of embryogeny or it may develop into a callus depending upon the plant species.

Irrespective of the early events in the division of microspore nucleus, there are two modes of androgenesis, the direct and indirect.

14.6. DIRECT ANDROGENESIS

In this, the microspore behaves like zygote and undergoes various stages of embryogeny simulating those *in vivo*. The embryos mostly at the globular stage are released from the exine and develop further. Finally, the cotyledons unfold and the plantlets emerge from the anthers.

Examples: *Atropa, Datura, Nicotiana*, *Petunia axillaries, Brassica campestris, Brassica napus.*

14.7. INDIRECT ANDROGENESIS

In contrast to the direct androgenesis, the microspores instead of undergoing embryogenesis, divide a few times to form callus which bursts through the anther wall. This mode of development is quite common and is usually caused by a complex media and in cases where the polarity seems to be disturbed. The callus either differentiates to form embryo or roots or shoots on the same medium or it has to be transferred to another medium. The callus derived plants are mostly undesirable as they exhibit genetic variations and polysomy.

Examples: Wheat, tomato, rice, barley.

14.8. CULTURE MEDIA AND NUTRITIONAL REQUIREMENTS

Basal media of White's, MS and Nitsch and Nitsch's with slight modification and the addition of growth regulators have been used for the culture of excised anthers. The normal level of sucrose is 2 to 4%. However, barley, tomato and wheat anthers have been observed to grow better on a media with 6-12% sucrose. This seems to be an osmotic effect rather than a need for high carbohydrate level.

The iron in the medium plays a very important role and is indispensable. Chelating agents such as Fe-EDTA, Fe-EDDHA is more efficient than ferric citrate.

Excised anthers require a much simpler medium than the isolated microspores. Anthers can be cultured on basal medium whereas, the microspores require higher amounts of nitrogen in the form of amino acids.

In some plants like *Atropa*, tobacco, etc, complete androgenesis can be realised on a simple media. But in most cases, a certain balance of auxins/cytokinins or complex additives such as casein hydrolysate, yeast extract, coconut milk and other extracts have to be incorporated in the medium.

Media rich in growth regulators encourage the proliferation of tissues other than the microspores and should be avoided because in such cases, mixed calli with cells of different ploidy levels are obtained. This would necessitate a laborious screening programme for the selection of haploid cells and plants.

Addition of charcoal in some cases has improved androgenesis, e.g., Tobacco. But it will increase the induction of diploid plants also. The stimulatory effect of charcoal is due to the regulation of both endogenous and exogenous level of growth regulators by absorption on to charcoal.

14.9. CULTURE OF ISOLATED POLLEN GRAINS/POLLEN CULTURE

Although anther culture has proved to be quite efficient for the induction of haploids, it has one main disadvantage:

- The plants not only originate from the pollen but also from various other parts of the anther with the result that a population of plants with various ploidy levels is obtained.

This difficulty can be circumvented by the culture of isolated pollen. Pollen culture involves the culture of pollen that has been removed from the anther. Pollen culture has been done in *Taxus brevifolia, Brassica oleraceae* and *Lycopersicon esculentum.*

The culture of pollen offers the following additional advantages:

- Uncontrolled effect of the anther wall and other associated tissue are eliminated and various factors governing androgenesis can be better regulated.
- The sequence of androgenesis can be observed starting from a single cell.
- Pollen is ideal for uptake, transformation and mutagenic studies as pollen may be evenly exposed to chemical or physical mutagens.

- Unlike single callus cells, pollen is transformed directly to an embryo, this would be most suitable for understanding the physiology and biochemistry of androgenesis.
- Higher yield of plants per anther could be expected.

14.10. METHODS OF POLLEN CULTURE

There are mainly three methods of culturing isolated pollen:

1. Hanging drop method
2. Nurse Culture technique
3. Pre-culture method

14.10.1. Hanging Drop Method of Pollen Culture

Kameya and Hinata (1970) first used this method to culture isolated pollen of *Brassica oleraceae and Brassica oleraceae* × *Brassica alboglabra*. This method involves placing a drop of medium containing 50-80 pollen grains on a cover glass, which is then inverted over a cavity slide and sealed with paraffin. Before inverting, the cover glass, a column of paraffin is raised in the centre so that when inverted, it touches the bottom of the cavity slide. This facilitates aeration as well as the movement of pollen when the slide is rotated. Cell clusters are formed from the isolated pollen on a medium containing coconut water.

14.10.2. Nurse Culture Technique

Sharp *et al.* (1972) induced the isolated pollen of tomato to form haploid callus. In this method, the anthers are placed horizontally on the top of basal medium within a French square container. A filter paper disc is placed over the intact anther and about 10 pollen grains (in suspension) are then placed on the filter paper disc. Clusters of green parenchymatous cells were formed on the filter paper discs in two weeks. These clones were observed to be haploid.

14.10.3. Pre Culture Method

Here the anthers are removed aseptically. After four days of incubation at 27°C the pollen are squeezed out of the anther into the liquid medium. After filtration the pollen are centrifuged to form a pellet. The pollen are washed twice, centrifuged again, and suspended in fresh media with the density of about 5×10^4 pollen/ml. Aliquots of 2 ml are then dispensed in thin layers in small Petri dishes or 25 ml Erlenmeyer flask or it can be grown in drop cultures. Here a drop of silicon is placed in the center of a small Petri dish and a cover slip is gently lowered onto the drop. Then a drop of 250 - 500 ml of the pollen suspension is pipetted onto the cover slip. The Petri plate is sealed with parafilm and incubated. The induced pollen undergoes normal androgenesis and eventually produces haploid plants. Certain refinements such as purification of pollen suspension to get viable microspores, serial culture of anthers in liquid medium enabling the embryogenic pollen to discharge in the medium in good quantities has been suggested.

14.11. PRECAUTIONS TO BE CONSIDERED DURING CULTURING OF POLLEN

- Growth and morphogenesis are adversely affected if the isolated pollens are not washed before culturing. Unwashed pollen either fail to grow or have a tendency towards callus formation.
- Cold treatment increases the frequency of androgenesis. The cold treatment inhibits spindle formation and in some cases, may cause abnormal first mitosis by triggering the pollen nucleus to form two similar looking nuclei. Cold treatment enhances the viability of cultured pollen and causes repression of the gametophytic differentiation, which results in higher frequency of androgenesis.
- Androgenesis in isolated pollen can be induced at various stages of development. But frequency of success is more by using anther taken at uninucleate to binucleate stage.
- Frequency of androgenesis can be considerably enhanced by raising the sucrose and myoinositol concentration in the medium.

14.12. CULTURE OF HAPLOID PROTOPLAST

The isolation of the protoplasts from pollen and pollen tetrads and their culture and regeneration is another approach in androgenic haploid production. Pollen grains of angiosperms are coated with sporopollenin, which is a polymer of carotenoids and carotenoid esters. This is dissolved only by KOH treatment, strong oxidizing solutions, certain organic bases and certain microorganisms. But, the yield of protoplasts has been limited, as no enzyme is known to digest exine. Better yield of protoplasts are obtained from pollen taken at earlier stages of development, *i.e.*, pollen mother cells (PMC) and tetrads. These are enclosed in callose wall, which is simpler compared to exine.

The procedure of isolation and culture of PMC and pollen tetrad involves the following steps:

- One anther is aseptically removed from the young flower bud and stained with acetocarmine to ascertain the stage of development.
- The basal end of the anther is cut with fine scalpel and the contents are squeezed out with a glass spatula. Excessive pressure during squeezing will press out diploid tapetal cells.

- The PMC and the tetrad emerge as a milky fluid and this is treated with an enzyme derived from snail intestine (0.75-1.0% helicase in 8-10% sucrose) for 30-45 minutes. 1 ml of enzyme is used per anther.
- After incubation, the enzyme mixture is carefully replaced by 10% sucrose solution and the protoplasts are allowed to settle.
- They are rinsed a couple of times with fresh medium and suspended in nutrient medium and culture.
- Centrifugation is not advised, as the protoplast yields that are too low will be further reduced.
- At places where they adhere, the tetrad protoplasts and developing microspores regenerate common walls giving rise to a clump of cells or so called microspore tissue, which is regenerated into haploid plants.

This method has yielded a limited success and their potential is not yet realized and commercialized.

14.13. FACTORS AFFECTING ANDROGENESIS

1. ***Genotype*:** Various species and cultivars exhibit different growth responses in culture. One of the reasons for failure in anther culture is that most of the workers restrict themselves to one cultivar and abandon the work if unsuccessful. It is highly desirable that general survey of various cultivars be undertaken. Further more, for such general survey, comparatively simpler media should be used as complex media rich in growth regulators tend to favour proliferation of somatic anther tissues giving rise to callus of various ploidy levels.

2. ***Physiological Condition and Age of the Plant*:** Flowers from relatively young plants at the beginning of the flowering season are more suitable. Plants grown under optimal growth conditions watered with mineral salt solutions periodically and relatively young plants should be used.

3. ***Stage of Pollen*:** Stage at which pollen are taken are most critical. For many species including Datura, tobacco, etc., the optimum stage is just before or just after the first pollen mitosis while the early binucleate stage is the most suitable for *Atropa belladonna* and *Nicotiana sylvestris*. In cereals, the best stage appears to be the mid-early or mid-uninucleate stage, i.e., before the first pollen mitosis. In tomato, the optimum stage is when the pollen mother cells are in meiosis I whereas, in *Brassica*, it is trinucleate or pollen shedding stage. In general, it is established that uninucleate microspores are more prone to experimental treatments for anther culture just before or during the first mitosis, while the culture of isolated pollen has yielded better results at the binucleate stage.

4. ***Thermal Trauma*:** Haploids have been observed to occur in nature and to be induced experimentally following some trauma. By temperature

manipulations, egg could be induced to develop parthenogenetically. Pollen also responds in similar fashion. Floral buds or entire plants subjected to hot or cold treatment or perhaps, alternate combinations of both prove to be useful in enhancing androgenesis.

The cold treatment effect is indirect. The low temperature retains the pollen viability longer, delays senescence and prevents the abortion of the pollen and thereby increases the number of available viable pollen, which is destined to form embryos.

Nemac observed that pollen grains of petalloid anthers of *Hyacinthus orientalis* on cold treatment, sometimes, assumed the eight nucleate condition resembling the embryo sac and lost their male potency. This is known as ***Nemac phenomenon***. This is of considerable interest.

5. ***Chemical Treatment***: Various chemicals are known to induce parthenogenesis. Colchicine treatment helped to obtain haploid plants in *Nicotiana langsdorfii* and beetroot. Ethrel is a chemical, which acts as a gametocide and cause male sterility. It is possible that multinucleate pollen isolated from ethrel treated anthers might induce to form embryos when culture and this way yield of haploids could be enhanced.

6. ***Composition of the Media***: Simple media with certain growth regulators are required for androgenesis. To promote direct embryogenesis, simple media with low levels of auxins are advisable. Complex media enriched with auxins such as 2, 4-D encourage the formation of callus, which causes genetic instability, and this should be avoided. Sugars are indispensable. They act as a source of carbon and are involved in osmoregulation. Addition of extract of androgenic anthers and potato tubers also stimulated androgenesis in tobacco. Supplementing the media with serine and glutamine improved the culture of isolated pollen.

7. ***Effect of Charcoal***: It absorbs inhibitory substances and regulates the level of growth substances.

14.14. STABILITY OF HAPLOIDS

One of the main problems with tissue culture, especially those of haploid origin is the maintenance of their genetic stability in long term culture. For their use in somatic cell genetics, it is essential that they remain haploid during early stages when the mutants are being selected. Parafluoro phenyl alanine (PFP) is a chemical used for such purpose. It is observed in tobacco cell cultures that PFP inhibited the growth of callus from diploid tissues whereas, the haploid tissue was unaffected. But in *Nicotiana sylvestris*, no preferential growth of haploid tissue was seen. So it was suggested that PFP acts as a selective inhibitor, which reduces the growth rate of haploid cells less than that of the cells of higher ploidy, but the response to PFP depends upon the genotype and the physiological state and age of the plant.

14.15. GYNOGENIC HAPLOID PRODUCTION

This involves haploid production utilizing the female parts of the plant. The spontaneous occurrence of haploids in nature usually takes place through parthenogenesis, but in some cases it can be induced experimental treatments especially by the application of inactivated pollen or distantly related pollen. Haploid parthenogenesis can be induced by culture of unpollinated ovaries or ovules. Hence, this method is also referred to as ovary and ovule culture.

Ovary and ovule cultures refer to the culture of unfertilized ovaries and ovules, respectively, to obtain haploid plants from egg or other haploid cells of the embryo sac. *In vitro* culture of unpollinated ovary and ovule has been tried in species where anther culture has given unsatisfactory results.

Raising of parthenogenetic haploid plants under *in vitro* conditions by gynogenesis was first demonstrated by San Noeum in 1976. She obtained gynogenic haploid by ovary culture of *Hordeum vulgare*. Haploids have also been obtained from unpollinated ovule cultures of *Gerbera jamesonii, Lilium* sp, and *Zea mays*. Induction of parthenogenetic haploid embryo after pollination by irradiated pollen was attempted in watermelon was attempted by Sari *et al.* (1994). They treated the pollen with gamma irradiation and such pollen was used for fertilizing the flowers. The embryos were then cultured *in vitro*. Success on similar lines has been reported in several other crops such as melon (*Cucumis melo*).

14.16. LIMITATIONS OF *IN VITRO* HAPLOID PRODUCTION

- Lack of development of techniques for quick production of large number of haploids.
- High cost of obtaining haploids and doubled haploids.
- Sometimes, deleterious mutations may be induced during *in vitro* phase.
- A non-random recovery of haploids may reduce the spectrum of variability recovered.
- The need for sophisticated tissue culture laboratory and dependable greenhouse.
- The need for specialized skills for carrying out various operations.

14.17. ANTHER CULTURE

Classic anther culture studies were done by Guha and Maheshwari in 1964 on *Datura* sp. students could refer to this protocol. But here, a protocol on *Nicotiana tabacum* is provided.

Objectives

- ➢ To induce callus and plantlet regeneration from anthers of *Nicotiana tabacum*

Equipment

- Petri dishes, Forceps, Dissecting needles, Scalpels
- Dissection microscope
- Laninar flow hood

Media and Reagents

Anther culture media : MS basal media
Root induction media : MS basal media with 1.0 ppm NAA
$HgCl_2$ (0.01%)
Surfactant (Tween-20)

Plant Material

Flower buds of *Nicotiana tobacum*

Procedure

1. Collect the flower buds of 21-23 mm size using forceps, as pollen. (Such buds would be in the late uninucleate or early binucleate stages).
2. Chill the buds at 7-8°C for 12 days in refrigerator.
3. Surface sterilize the buds in a Petri dish with 0.01% of $HgCl_2$ with a few drops of Tween-20 for 10 minutes.
4. Rinse the buds 3-4 times in sterile distilled water in a sterile environment.
5. Carefully tease open the buds, remove the anthers using forceps and dissecting needle.
6. Discard the filaments to avoid their callusing.
7. Place the anthers on the anther culture medium and incubate the cultures at 25°C in dark.
8. After 2 weeks, search for young embryos by gently bursting the anthers on a microscope slide and observe under microscope. The growth of the young embryos to complete plantlets can be observed after 4-5 weeks of culture.
9. Separate the plantlets with forceps and transfer them to root inducing media when they are about 1 cm in size. When the plantlets are about 5 cm in size they could be transferred to *ex vitro* conditions.

14.18. OVARY AND OVULE CULTURE

Objective

- ➢ To induce haploid plantlets from ovary and ovules of flowers of tomato.

Equipment

- Petridishes, forceps, scalpel

Media and Reagents

- Nitsch's media
- Mercuric Chloride (0.01%) or Sodium hypochlorite (0.8%)

Plant Material

Flowers of *Lycopersicon esculentum*

Procedure

1. Collect flowers of tomato along with the pedicel.
2. Surface sterilize the flowers with mercuric chloride under strerile environment.
3. Rinse with sterile distilled water to remove traces of the sterilent present.
4. Remove the calyx, petals, etc., and retain only the ovary.
5. Sterilize the material with sodium hypochlorite and rinse again with sterile distilled water.
6. Culture the ovary as such on the culture media. Also, cut the ovary into small pieces and culture them to induce plantlet regeneration from the ovules.
7. Incubate the material at 25°C under dark conditions.

14.19. PROBLEMS ASSOCIATED WITH *IN VITRO* HAPLOID PRODUCTION

1. Often anthers fail to grow *in vitro* or the initial growth may be followed by abortion of the embryos.
2. The tissue or calli developed generally comprises a chimera of diploid, tetraploids and haploid cells.
3. Low economic viability due to low success rate.
4. Doubling of chromosome number of the haploids is time consuming and may not always result in the production of a homozygote.

15

EMBRYO CULTURE

The embryo represents the beginning of the new sporophytic generation of the plant. In virtually all the higher cryptogams, the embryo has its origin in the fertilized egg or the zygote, which is generally buried within a privileged and relatively in accessible location in the female gametophyte. Embryo of different developmental stages formed within the female gametophyte through the normal sexual processes, can be separated with relative ease from the bulk of the maternal tissues and cultured *in vitro* under aseptic conditions in media of known chemical composition.

15.1. EMBRYO CULTURE AND ITS APPLICATIONS

Embryo culture involves the isolation and germination *in vitro* of fertilized or unfertilized seed embryos. Genetic variability for agronomic and horticultural traits is the basis for crop improvement, which in turn will ensure a stable agricultural economy and reliable food supply. One of the most diverse sources of variability is that of intergeneric hybridization. Although the phylogenic distances between genera are much greater than species differences, the developments over the past decade have permitted the bridging of such gaps. These developments mainly include the discovery of genetic crossing mechanisms and embryo culture techniques. The technique of embryo culture has been used for a number of different purposes including embryo rescue after intergeneric crosses, speeding up the breeding cycle by culturing *in vitro* a slow developing embryo breaking the dormancy of hard-to-germinate seeds, production of a viable plant after an enforced self-fertilization and in those instances where mature seed is not obtainable for a species because seed is shed and lost suddenly when ripe or developing seed is attacked by pests before ripening.

Embryo rescue technique is a much sought after technique of embryo and species exhibiting post-zygotic in compatibility, in hybrid crosses leading to embryo abortions. This has been successfully achieved in sepcies such as grapes, *Prunus* (Arora *et al.* etc., 2003)

15.2. EXCISION OF EMBRYOS

Seeds, ovules or fruits are surface sterilized before embryo removal. The embryo itself needs no sterilization provided aseptic conditions are strictly maintained during its excision and transfer to the medium. To ease the dissection of the embryo, hard seeds are soaked in water to soften them, but if softening takes more than a few hours, it is advisable to re-sterilize the seed afterwards. A dissecting microscope may be necessary to excise the embryos from small seeds, as it is particularly important that the embryo should not be damaged.

15.3. TYPES OF EMBRYO CULTURE

According to Pierik (1989), there are two types of embryo culture:

15.3.1. Culture of Immature Embryos

- This type of embryo culture is mainly used to grow immature embryos originating from unripe hybrid seeds, which fail to germinate. A complex nutrient media is required in this case.

15.3.2. Culture of Mature Embryos

- Mature embryos are excised from ripe seeds and cultured. The main purpose is to overcome inhibitory factors during seeds germination. This type of culture is relatively more easy. Fig. 9. illustrates the technique of embryo culture.

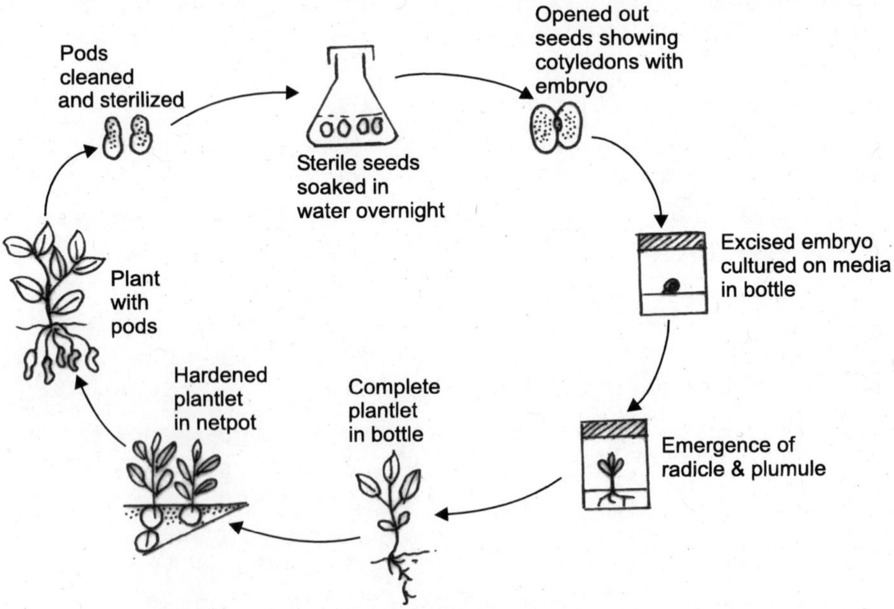

Fig. 9: Diagrammatic representation of embryo culture

Immature embryos (proembryos) are frequently more easily cultured than mature ones, however, their dissection requires much skill. Culture of proembryos also requires more complex media. In general, mature embryos require only inorganic salts supplemented with sucrose, whereas, immature embryos have an additional requirement for vitamins, amino acids, growth regulators and sometimes coconut milk or some other endosperm extract.

Some of the horticultural crops where embryo culture has been successfully done are *Abelmoschus* sp., *Allium cepa, Anacardium occidentale* and *Cocos nucifera* (mature embryo).

Objective

➢ To standardize the technique of embryo culture to be adapted in plant species of interest.

Facilities and Equipment

- Laminar airflow hood
- Stereomicroscope with illuminator for embryo excision
- Sterile culture tubes, Petri dishes
- Sterile forceps, scalpels

Media and Reagents

- MS basal media
- Mercuric chloride (0.5%)
- Sterile water

Procedure

(a) Excision of the Embryos

1. Soak hard-coated seeds in water for a few hours to a few days to loosen the seed coat and facilitate its easy removal.
2. Surface sterilize the seeds with a suitable sterilent for 5-10 minutes and rinse in sterile distilled water. Keep the seeds in sterile water until dissection.
3. Dissect the seeds on a sheet of sterile filter paper under a stereomicroscope in a laminar flow hood. Begin by locating the micropylar-plug indentation at the less pointed end of the seed. The embryo can be located very close to the micropylar-plug.
4. With a sterile scalpel blade, make a longitudinal incision through three-fourths of the median line with the cut end pointing towards the micropylar-plug. The remaining one-fourth of the uncut portion can be split open.
5. The excision of an immature embryo embedded in liquid endosperm involves an incision at the micropylar end of the young ovule and the

application of pressure at the opposite end to force the embryo out through the incision opening. Take care that the pressure does not injure the fragile embryonic tissue.

6. With a sterile dissection needle, transfer the embryo onto the medium.
7. Incubate the cultures in darkness for a week and then in light thereafter.
8. Once the plantlet has put forth four or more leaves, it can be planted *ex vitro*.

Schedule of Observations

- Take weekly data on *in vitro* embryonic development (percentage of embryos at heart, torpedo and mature stages) during the first month of incubation.
- Record the percentage of embryos showing developmental abnormalities.

15.4. COMMON PROBLEMS ENCOUNTERED

Growth suppression/cessation: Excised embryos may stay quiescent in culture, but healthy in appearance for months without any sign of growth. This may be due to the embryos entering a secondary dormancy for which no dormancy breaking procedure has yet been discovered. Small amounts of toxic compounds released during autoclaving of plastic containers or caps in culture medium may also result in growth suppression or cessation of the embryos without showing cell death or embryo discolouration.

Precocious Germination: Excised immature embryos on a nutrient medium, tend to bypass the stage of dormancy and cease to undergo the linear embryogenic mode of development. These embryos develop into weak seedlings. Casein hydrolysate, an amino acid complex if used as an additive in the media, will promote embryogenic development and prevents precocious germination. High sucrose in the mediun also prevents precocious germination. Other exogenous factors that suppress precocious germination are reduced oxygen tension, elevated temperature and high light intensity.

16
ENDOSPERM CULTURE

Endosperm tissues are the nutritive tissues found in all the angiosperms with a few exceptions (Orchidaceae, Podostemaceae and Trapaceae). It may be consumed wholly by the developing embryo or it may persist in mature seed as a massive tissue and store reserve food materials in diverse forms. Endosperm is a homogenous mass of parenchymatous tissue lacking any organogenic or vascular differentiation. It is triploid in nature, having been derived from the fusion product of three haploid nuclei (two from the female gametophyte and one from the male gametophyte). Embryologists regard endosperm as the second embryo modified to serve as the nutritive tissue for the zygotic embryo.

The stage of seed development at which the endosperm can be used to initiate cultures depends on several considerations. In plants with non- endospermous seeds (e.g., apple and citrus), only immature seeds can provide the necessary explants. In most of the dicotyledonous plants, endosperm tissue from immature as well as mature dried seeds has been used to initiate cultures. In some plants, the association of the embryo is essential for the initial stages of dedifferentiation of the cells of mature endosperm. In such cases, the entire decoated seeds are used as the explant. Once endosperm cells have started dividing, the embryo is removed and the endosperm tissue is transferred to fresh medium for further growth.

Lampe and Mills (1933) are credited for the pioneering attempts to culture immature endosperm of maize and noted a slight proliferation of endosperm layers adjacent to the embryo. Of late, the potential applications of endosperm culture for triploid production in relation to crop improvement have been recognized. The demonstration of the totipotency of triploid cells of endosperm has created a new approach to the production of a large number of triploid plants in one step. Where seed is not commercially important or seedlessness is desirable as in some fruit trees, and where plants can be propagated vegetatively, the induction of triploidy may prove important for the improvement of these crops. Some of the economically important plants where triploid plants of endosperm origin have been developed are *Actinidia chinensis, Citrus grandis,*

Pyrus malus and Santalum album. Triploids are also useful for obtaining trisomic lines for genetic mapping.

16.1. APPLICATIONS OF ENDOSPERM CULTURE IN HORTICULTURE

- Advances made in the technique of endosperm culture have enabled the production of triploid plants. Triploid plants are self-sterile and usually seedless. This trait increases the edibility of fruits and is desirable in crops such as apple, banana, mulberry, grape, mango, watermelon, etc.
- Triploid plants of *Petunia axillaris* were found to be more vigorous than the parents.
- Triploid production by conventional technique is laborious and may be unsuccessful. In such cases, endosperm culture is the only feasible technique.
- Endosperm provides an experimental system for studying biosynthesis and metabolism of natural products, e.g., Endosperm callus of coffee was shown to synthesize caffeine. (Fig. 10)

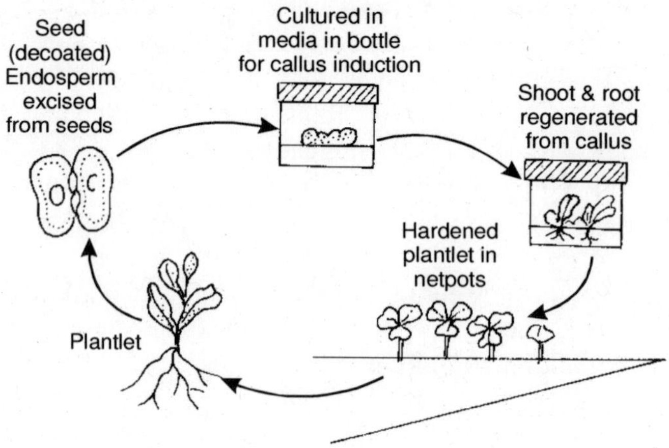

Fig. 10: Diagrammatic representation of endosperm culture

16.2. OBJECTIVE

- To induce triploid plantlets from endosperm tissue.

16.3. EQUIPMENTS

- Laminar airflow hood
- Stereomicroscope with illuminator for excision of endosperm

- Sterile culture tubes, Petri dishes
- Sterile forceps, scalpels

16.4. MEDIA AND REAGENTS

- Basal White's medium or MS medium (For the establishment of endosperm cultures of dicotyledonous plants, the presence of an auxin (2,4-D or IAA), a cytokinin (Kinetin or BAP) and a rich source of organic nitrogen (yeast extract or casein hydrolysate) in the basal medium is essential).
- MS media with a cytokinin and low concentration of an auxin (IAA/IBA)
- Distilled water
- 70% alcohol
- Sodium hypochlorite/Mercuric chloride
- Agar (0.8%)
- Sterile water

16.5. PROCEDURE

(a) Explant Preparation

- ***Mature endosperm*:** In plants having seeds with massive endosperm, the decoated seeds are directly surface sterilized with a suitable disinfectant. After 2 to 3 washings in sterile distilled water, they can be placed on the medium.
- ***Immature endosperm*:** Entire seeds are surface sterilized. The endosperm tissue is carefully excised under aseptic conditions using a stereoscopic microscope and then placed on the basal medium for callus induction.

Note: Half-split endosperm pieces planted with their cut surface in contact with the medium are more responsive than cultures with their cut surface of endosperm away from the medium.

(b) Shoot Bud Differentiation

- The calli are transferred to MS media supplemented with cytokinin and low concentration of auxin for the differentiation of shoot buds from the endosperm tissue.
- To induce embryo formation, the calli may be transferred to a basal medium.

Note: In some species, plantlet formation from endosperm callus may follow the embryogenic mode of development.

16.6. PROBLEMS ENCOUNTERED

In plants where entire decoated seed is used as explant, the embryo may produce callus along with that from endosperm. This can be avoided by culturing excised endosperm pieces presoaked in GA_3 solution. GA_3 treatment helps to replace the need for embryo factor.

Endosperm culture has been fairly successful in few plant species. One such example is mulberry. However, our studies at UAS, Bangalore, have not been very encouraging. Though endosperm tissues have given rise to callus and shoot organogenesis have been meek this has not enabled us to cytogenically prove triploidy nature of the tissue. Only increased number of chromosomes from cytological studies of the callus tissue could be as ascertained. (Gopalswamy, Nutan and Sathyanarayana, 2004).

17

Culturing of Roots *In Vitro*

One of the pioneering events in the development of tissue culture techniques was the successful establishment of actively growing clones of tomato roots reported by P.R. White in 1934. In this classical research, White cultured tomato roots through 160 successive passages. The discovery that roots could be grown aseptically in this isolated way was the first significant development of modern tissue culture science. Although root hair cultures (using *Agrobacterium rhizogenes*) are more popular today, this technique still maintains a certain level of interest in root biology.

Root cultures can be established from root tips taken from primary or lateral roots of many plants. Suitable explants are small sections of aseptic roots bearing a primary or lateral root meristem. These explants may be obtained from surface sterilized seeds germinated in aseptic conditions. If the small root meristems continue normal growth on a suitable medium, they produce a root system consisting only of primary and lateral roots. No organized shoot buds will be formed.

Isolated plant roots can usually be cultured on relatively simple media such as White's containing 2% sucrose. Liquid media are preferable, as growth in or in a solid media is slower.

Isolated root cultures have been employed for a number of different research purposes. Cultures of this kind may also be used to culture mycorrhizal fungi aseptically, and to study the process of root nodulation with nitrogen fixing *Rhizobium* bacteria in leguminous plants. Isolated root cultures do not feature in current micropropagation techniques, although shoots can be regenerated from root segments of some species. It has, however, been suggested that root cultures could afford one means of germplasm storage.

Examples of horticultural crops where root cultures have been employed: *Cucumis melo, Pisum sativum*, etc.

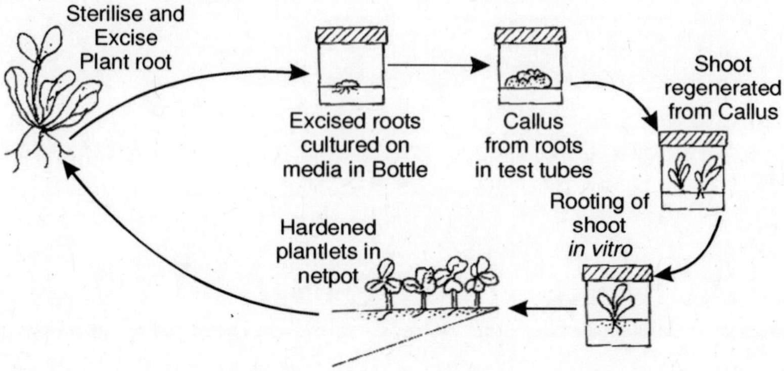

Fig. 11: Diagrammatic representation of in vitro culture of roots

17.1. OBJECTIVE

➢ To establish root culture of pea (*Pisum sativum*)

17.2. MATERIALS REQUIRED

- Sterile Petri dishes
- Conical flasks
- Scalpels
- Forceps

17.3. MEDIA AND REAGENTS

- White's media supplemented with 2,4-D at 0.01 mg/L and 1.0 mg/L
- Ethanol (95%)
- Tween-20
- Sterile distilled water

17.4. PROCEDURE

1. Place the pea seeds in a conical flask, pour 200 ml of 95% ethanol over the seeds and shake for 10 seconds, then pour off the ethanol.
2. Pour 200 ml of Tween-20 solution over the seeds and shake the flask gently for one minute.
3. Examine the seeds individually and discard those that float or appear to have a translucent tissue beneath the seed coat.
4. Sterilize the remaining seeds by pouring 200 ml of Chlorox solution (disinfectant) over the seed and disinfect for 20 minutes under aseptic conditions. Following sterilization, wash the seeds three times in sterile distilled water using about 200 ml per rinse.

5. Pour 20 ml of sterile distilled water into each of the ten Petri dishes and place two seeds in each dish. Incubate the plates in the dark at 25°C for 48h.
6. Prepare White's basal medium with the growth regulators at two different concentrations as given earlier and sterilize it. Under aseptic conditions, dispense 20 ml of the medium into the each of the ten Petri dishes. Repeat the procedure for both the media with growth regulators at different concentrations.

Plate 46: Culture of excised root tips in potato (in MS media with NAA 0.05 mg/L)

7. When the primary root of each pea seedling reaches 20-25 mm, excise the terminal 10 mm and transfer the tip to a plate containing similar medium as earlier. Inoculate two tips per plate.
8. Incubate all the cultures in the dark. (Plate 46 shows *in vitro* cultured root tips of potato)

18

Protoplast Isolation and Culture

A protoplast is the living part of a plant cell, consisting of cytoplasm and nucleus with the cell wall removed. Methods of isolating plant protoplasts have made significant advances during recent years and protoplasts have now been obtained from whole plant organs or tissue cultures of many different species. Techniques have also been developed to enable isolated protoplasts to be cultured so that, by growth of a cell wall followed by cell division, they may give rise to callus and thence to whole plants. Protoplast culture therefore provides one route whereby plants can be multiplied, but it is not yet used for routine micropropagation work, and the number of species in which plant regeneration has been achieved is still relatively small.

At present, isolated protoplasts are used chiefly in research into plant virus infections, ways to modify the genetic information of plant cells, including the creation of plant cell hybrids through protoplast fusion. However, genetically modified cells will be only of general practical value if whole plants having new genetic constitution can be regenerated from them. The ability to propagate plants from protoplast cultures is therefore of vital importance to the success of such genetic engineering projects in plant science.

18.1. PROTOPLAST ISOLATION

18.1.1. Sources of Material

Protoplasts can be isolated from almost all parts viz., roots, leaves, fruits, tubers, root nodules, endosperms, crown gall tissues, pollen mother cells, pollens, pollen tetrads and cells of callus tissue grown *in vitro*. The most convenient and popular source of plant protoplasts is the leaf, as the mesophyll cells are loosely arranged and the chemicals used for isolation have easy access to the cells. In the case of plants grown in greenhouses or in growth-controlled rooms, better results have been obtained. *In vitro* grown shoots have also been used as a source of plant protoplasts. Some species of cereal crops show difficulty in isolating culturable protoplast from leaf cells, the alternate source in such cases is cell suspension cultures.

18.1.2. Methods of Isolation

18.1.2.1. Use of Osmoticum

Tissue from an entire plant to be used for protoplast separation is first surface sterilized. During the isolation process, one problem is the possibility of bursting of the isolated protoplasts due to their osmotic fragility. To prevent bursting, the use of a suitable osmotic stabilizer in the solution is advisable. The freshly isolated protoplasts in a proper osmolarity appear spherical. The protoplasts are more stable in a slightly hypertonic solution rather than isotonic solution.

For successful isolation, it has been found essential to cause the protoplast to contract away from the cell wall, to which, under normal circumstances, it is tightly adhered and connected. Contractions are achieved by plasmolysis of the cells with solutions of salts such as potassium chlorate, magnesium sulphate or with sugars or sugar alcohols (particularly mannitol). In case of mechanical isolation method, salt solutions ($CaCl_2$-50 to 100mM L^{-1}) are predominantly used as osmotic stabilizers while in enzymatic methods, sugars or sugar alcohols are used. These osmotica must be of sufficient concentration to cause shrinkage of the protoplasm, but of insufficient strength to cause either damage to the cells or spontaneous protoplast fusion upon isolation. Plasmolysis helps to protect the protoplast when the cell wall is ruptured during mechanical separation and also appears to make it more resistant to the toxic effects of enzymes used for cell wall digestion.

There are several different methods by which protoplasts may be isolated:

(*a*) By mechanically cutting or breaking open the cell wall.

(*b*) By digesting away the cell wall with enzymes.

(*c*) By a combination of mechanical and enzymatic separation.

18.1.2.2. Mechanical Isolation of Protoplasts

In the past, protoplast have been isolated successfully from pieces of sectioned plant material but only a very small number of intact undamaged protoplasts were obtained by this method.

1. The leaf material is taken in salt solution and sucrose (1%) and ground in a homogenizer until all the leaf pieces are broken up. The cells separate clearly without causing extensive cell damage.
2. The cells are filtered and the debris is removed and the resulting cells are used for culture.

This method is rarely used, as it is an extremely tedious process that results in the yield of only very small number of protoplasts. Mechanical methods have now been almost completely replaced by enzymatic isolation techniques.

18.1.2.3. Enzymatic Isolation

Enzymatic method is most often used in isolation of protoplasts from various plant parts. Commercially available enzymes used for protoplast isolation are each partially purified mixture of several enzymes from a fungal or bacterial source, and have pectinase, cellulase or hemicellulase activity. Enzymatic methods generate very large number of protoplasts in comparison to mechanical methods with very less damage to the cells and lesser shrinkage to the protoplasts. However, in some instances, the enzymes have deleterious effects on plant metabolism.

Enzymatic isolation can be performed in a one-step (direct method) or two-step (sequential method) isolation process.

1. The leaf or other tissue segment is macerated with a macerozyme (a pectinase) in 13% mannitol to isolate the cells.
2. The isolated cells are purified by filtration through a nylon mesh.
3. These isolated cells are incubated in 2% cellulase for about 90 minutes for isolation of protoplasts.
4. In one-step process, both the enzymes (pectinase + cellulase) are used simultaneously. In this method, 2 to 3 hours at 20 to 22°C is adequate to release significant numbers of protoplasts.

The technique of protoplast isolation culture and regeneration of plants have been illustrated in Fig. 12.

18.1.3. Protocol for Protoplast Isolation from Leaves

The physiological state of the leaves is extremely important for the successful isolation of protoplasts from the leaves. The leaves normally chosen for experimental purposes should be fully expanded and from young plants. The lower epidermis is removed and is used to isolate epidermal and guard cell protoplasts. The remaining peeled leaf is used for isolation of mesophyll protoplasts. The isolation of protoplasts involves three main steps:

18.1.3.1. Preparation of the Explant

1. Surface sterilize fully expanded young leaves by dipping them in 70% alcohol for one minute and then washing in 2% sodium hypochlorite solution for 20 to 30 minutes.
2. Wash the leaves three times in sterile distilled water to remove traces of the sterilent.
3. The subsequent steps to isolate the protoplasts are taken up under aseptic conditions.
4. Carefully peel off lower epidermis of the sterilized leaves and cut the stripped leaves into small pieces. (Peeling off process is easier with flaccid leaves than with turgid leaves).

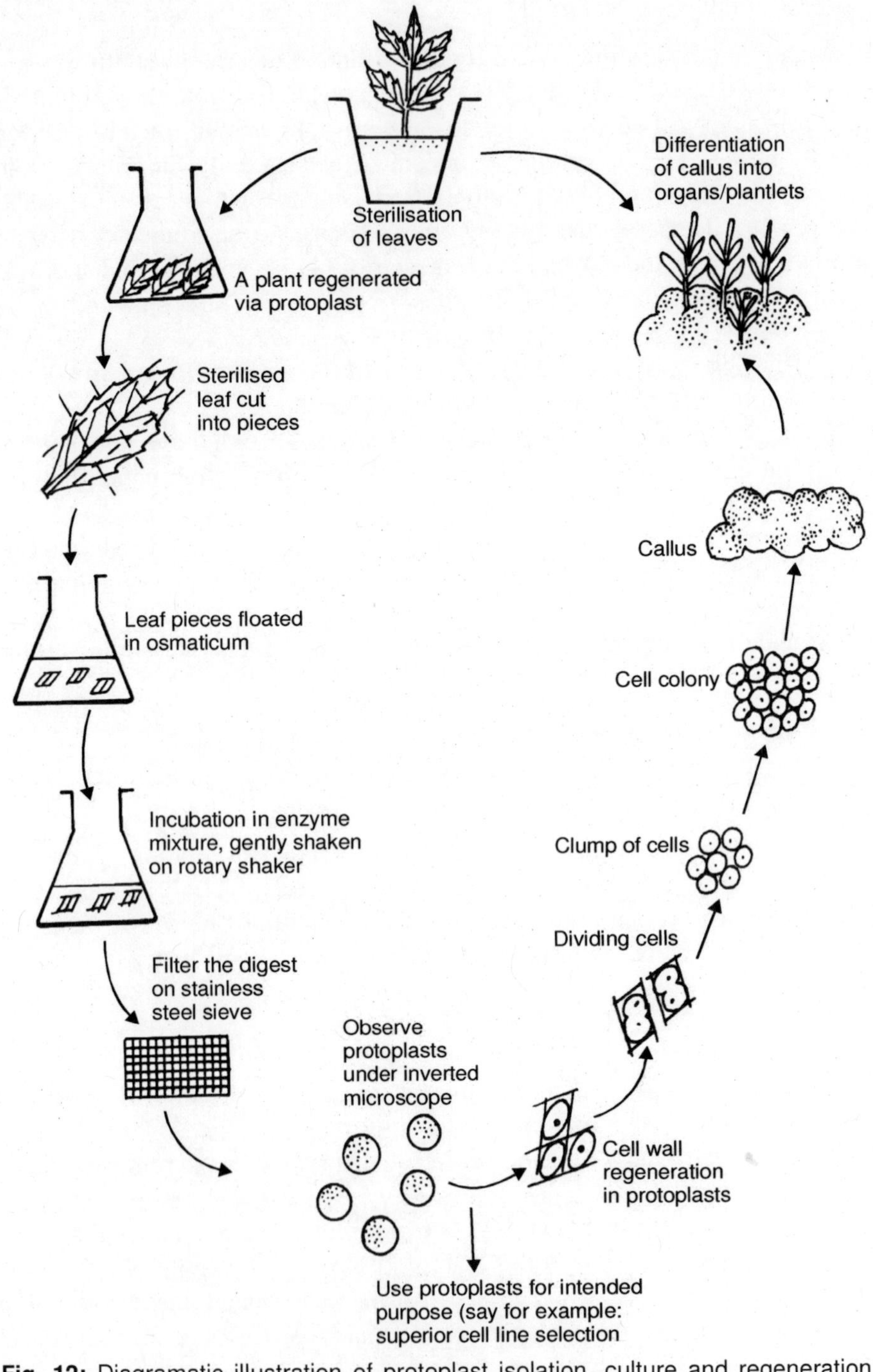

Fig. 12: Diagramatic illustration of protoplast isolation, culture and regeneration of plants from protoplasts

18.1.3.2. Protoplast Isolation by Enzymatic Method

The protoplasts can be isolated by direct or sequential method.

Direct Method

1. Place the peeled leaves with lower surface down, into 20 ml solution of 13% mannitol and inorganic salts contained in Petri dishes for one hour.
2. Pipette out the mannitol-inorganic salt solution and add an enzyme mixture (20 ml of 0.5% macerozyme and 2% cellulase in 13% sorbitol or mannitol at pH 5.4), incubate overnight (15 to 18h) at 20 to 22°C.
3. Gently agitate the Petri dishes to facilitate the release of protoplasts and leave it for a few minutes so that the protoplasts will settle down.
4. Transfer the protoplast-enzyme mixture to centrifuge tubes and centrifuge the material at 100 × g for one minute. Repeat this process 2 to 3 times and wash the protoplasts with 13% sorbitol solution. For the final washing, use 20% sucrose solution in place of sorbitol and centrifuge at 200 × g for one minute.
5. Pipette out the floating protoplasts leaving behind the remains of mesophyll cells.

Sequential Method

Isolation of the Cells

1. Place the peeled leaf segments in a Petri dish containing pectinase enzyme (0.5% macerozyme, 0.3% potassium dextran sulphate in 13% mannitol at pH 5.8) and agitate them slowly.
2. After 15 to 20 minutes, replace the enzyme mixture by fresh one and again incubate for one hour.
3. Filter the isolated cells through a nylon mesh and then centrifuge at 100 × *g* for a minute.

Isolation of Protoplasts

1. Treat the isolated cells with another enzyme mixture containing 2% cellulase in a 13% solution of mannitol at pH 5.4 for 90 minutes at 30°C and then centrifuge at 100 × g for a minute.
2. Decant off the supernatant and wash the protoplasts three times with mannitol.
3. Finally, clean the protoplasts with 20 to 25% sucrose solution.

18.1.3.3. Purification of the Protoplasts

1. Gently transfer the protoplasts, enzyme solution and the undigested tissues onto a 64 or 44 μm pore-size filter inside a sterile funnel.

2. Collect the filtrate in a centrifuge tube and centrifuge at 500 × *g* for 5 to 10 minutes.
3. Remove the enzyme solution carefully with a sterile pipette.
4. Add 12 ml of 0.4M sucrose to the pellet, and gently disperse and resuspend the protoplast in the solution.
5. When the protoplasts are completely dispersed, add carefully by layering over the sucrose solution a mixture of 0.4M sorbitol, 10mM $CaCl_2.2H_2O$, at pH 5.8 and centrifuge at 250 *g* for 6 minutes.
6. To wash the protoplasts, collect the protoplasts carefully with a pipette from the interphase of the two solutions and transfer to 4 ml of salt solution containing 0.2%w/v $CaCl_2$ and 2.5% KCl at pH 6.0. Gently suspend and centrifuge at low sped to collect the protoplasts.

18.2. TESTING THE VIABILITY OF THE PROTOPLASTS

In early days, the viability of the protoplasts was determined by observing its Cytoplasmic streaming, exclusion of Evans Blue dye, changes in protoplast size induced by changes in the level of osmoticum and the measure of photosynthetic and respiratory activity. However, most of these methods have proved to be rather unreliable for determining protoplast viability.

The most commonly used method for determination of viability is the use of fluorescein diacetate (FDA). As the FDA accumulates within the plasma membrane, viable protoplasts fluoresce green or white. (Protoplast preparation treated with FDA (normally 0.01%) should be observed within 5 to 15 minutes, because after this period, FDA dissociates from the membrane). Another stain commonly used is phenosafranine (0.01%). This stain detects dead protoplasts, which turn red in the presence of the dye. Viable protoplasts remain unstained even after 2h of staining.

18.3. CULTURE OF PROTOPLASTS

Isolated protoplasts can be cultured in several ways which includes in liquid culture medium, agar embedding technique, as hanging drops or in thin layers of liquid over a layer or nutrient medium solidified with agar. Of these techniques, agar-embedding technique is most commonly followed.

Isolated plant protoplasts are very fragile and particularly liable to either physical or chemical damage. Hence, if they are suspended in a liquid medium, it must not be agitated, and the high osmotic potential of the medium in which isolation was carried out must be temporarily maintained. As growth depends on adequate aeration, protoplasts are usually cultured in very shallow containers of liquid or solid media and a fairly high plating densities (5×10^4 to 10^5 protoplasts per ml) is essential possibly because endogenous chemicals are liable to leak away from such unprotected cells. To promote growth, it may also be beneficial to add to the medium supplementary chemicals and growth factors not normally required for the culture of intact cells.

18.3.1. Agar Embedded Culture

In this method, protoplasts are embedded in solid agar medium and is often referred to as *plating of protoplasts*. The plated protoplast can be handled very easily and the agar medium gives good support to the protoplasts. The isolated protoplasts may be allowed to regenerate a new cell wall in liquid culture before embedding or may be embedded directly after isolation. The protoplasts are mixed with 1.0% agar culture medium and maintained at 40 to 45°C. Small amount of this culture medium is then poured into sterile Petri dishes.

18.3.2. Microchambers

Here, a chamber is constructed of a small plastic ring with a coverslip on top. The joints of the system are sealed with mineral oil. The simplest form is to hold a droplet between two cover slips with a ring of mineral oil around to form a seal. The mineral oil prevents the vapourization of the media, which may result in the rise in osmotic concentration of the media.

18.3.3. Hanging Drop Culture Technique

A drop of the solution with the protoplasts is placed in the depression of a specially prepared hanging drop-slide, which is then under sealed with a cover slip and oil. Alternatively, the drops may be placed on the lid of a Petri dish with mannitol solution to maintain the humidity. The dish is sealed with parafilm.

18.4. REGENERATION OF CELL WALL BY THE CULTURED PROTOPLASTS

The capability of plant protoplasts to divide appears to be closely related to their ability to form a cell wall. Under favourable circumstances, formation of a cell wall seems to occur as soon as protoplasts are removed from hydrolyzing enzyme preparations. The first signs of cellulose deposition can be detected after about 16h in culture medium. The newly built cell wall can be observed by either plasmolyzing the protoplast by transferring it into a hypertonic solution, or by staining the cell wall with calcofluor white fluorescent stain (0.1%). Calcofluor binds to the newly formed wall material and would fluoresce when observed using a mercury vapour lamp. Once cell wall formation is initiated, the concentration of osmoticum is reduced to favour cell growth. This is readily accomplished in a liquid medium, but where protoplasts have been plated on a solidified agar medium, it will be necessary to transfer the cells on blocks of agar.

18.5. REGENERATION OF PLANTLETS

Once the cell wall is formed, the regenerated plant cell generally increases in size and may divide in 3-5 days. Further cell divisions may give rise to small colonies of intact cells. After 2-3 weeks, macroscopic colonies are formed, which can be transferred to an osmotic free medium to develop a callus. Regeneration of plantlets can either takes place by organogenesis or by embryogenesis. Organogenesis results from the callus and not directly from a single cell. Alternatively, embryoids may be initiated from superficial clumps of cells with dense cytoplasm and large starch grains.

Crops where plant regeneration from cultured protoplasts have been achieved:

Lactuca sativa, Cichorium intybus, Brassica campestris, Brassica oleraceae var. *capitata, Cucumis sativus, Manihot esculenta, Asparagus officinalis, Citrus aurantifolia, Citrus grandis, Citrus limon, Citrus medica, Citrus paradisi, Citrus reticulata, Citrus sinensis, Atropa belladonna, Capsicum annum, Datura metel, Datura innoxia, Hyoscyamus muticus, Lycopersicon esculentum, Petunia hybrida, Petunia inflata, Petunia parodii, Petunia parviflora, Solanum tuberosum, Solanum melongena, Solanum khasianum, Solanum viarum, Solanum torvum, Daucus carota, Foeniculum vulgare.*

18.6. PROTOCOL FOR PROTOPLAST ISOLATION FROM CARROT (DAUCUS CAROTA) (Subramanya and Sathyanarayana, 2002)

Explant Preparation

- Roots of carrot serve as the explant. Wash the explant material thoroughly under running water to remove the dust and soil particles.
- Cut the roots into circular discs of 0.5 cm thickness.
- Keep the discs in Tween-20 (Surfactant) for 10 minutes.
- Wash the discs in distilled water followed by treatment with sodium hypochlorite (0.25%) in a sterile environment.
- Rinse the material in sterile water several times and then cut into small pieces of 0.25 cm length.

Protoplast Isolation

- Plasmolyze 500 mg of the explant material in 20 ml of 12% mannitol solution for two hours.

- Cut the material into small pieces and incubate in 5 ml of cellulase (0.75%) and macerozyme (0.4%) (for digestion of the cell wall) with 5mM sorbitol and incubate for 16 hours under dark conditions.

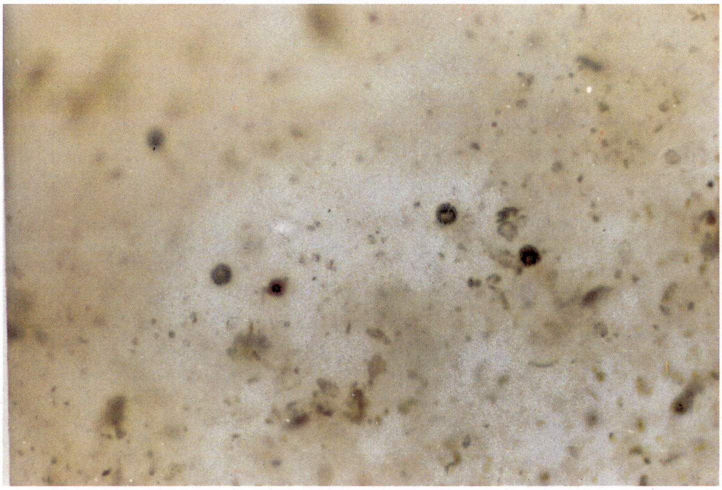

Plate 47: Protoplast isolated from carrot

18.7. ISOLATION OF PROTOPLASTS FROM LEAVES OF CHILLI (*CAPSICUM ANNUUM* L. VAR *ACUMINATUM*) (Deepu Mathew and Sathyanarayana, 2000)

Collection of Explants

- Collect leaves from 4-week old seedlings of chilli.
- Wash the explants thoroughly in running water for 15 minutes to remove the adhering dirt.
- Rinse the explants in distilled water containing two drops of Tween-20 for 10 minutes followed by washing in sterile water.

Isolation of Protoplasts

- Cut the leaves into 0.1-0.2 cm strips using a sharp stainless steel blade.
- Float the material on osmoticum (9.1 g of 0.5M mannitol and 140 mg of 10mM calcium chloride dissolved in 100 ml of double distilled water). Transfer 0.6 g of leaf pieces to 100 ml conical flasks with 10ml of osmoticum.
- Incubate the material in enzyme mixture (1.5% celluclast + 0.5% Pectinex) overnight at 26°C in the dark.
- 4-6 h before the completion of digestion, gently shake the mixture on a rotary shaker (<10 rpm).

- Filter the digest through a 100 m pore size stainless steel sieve.
- Mount one drop of the filtrate on a slide and observe using an inverted microscope. (This gives an idea about the efficiency of digestion).

Purification of the Protoplasts

- Distribute the filtrate into screw cap tubes and centrifuge for 10 minutes at about 600 rpm.
- The protoplasts collect as a pellet at the bottom of the tube.
- Decant the enzyme solution using a pipette without disturbing the pellet.
- Suspend the pellet of protoplasts in wash medium (osmoticum).
- Repeat centrifugation and re-suspend in fresh medium twice so as to remove the traces of enzymes.
- Further suspend the protoplasts in 2 ml of wash medium.
- Layer 2 ml of the concentrated protoplast suspension on top of about 9 ml of floating medium (17% sucrose) in screw cap tubes and centrifuge for 10 minutes at 600 rpm.
- The healthy protoplasts will be concentrated as a band at the interface of the wash medium and floating medium.
- Carefully collect this with a pipette to a screw cap tube. (Plate 48 shows protoplasts isolated from chilli leaves.

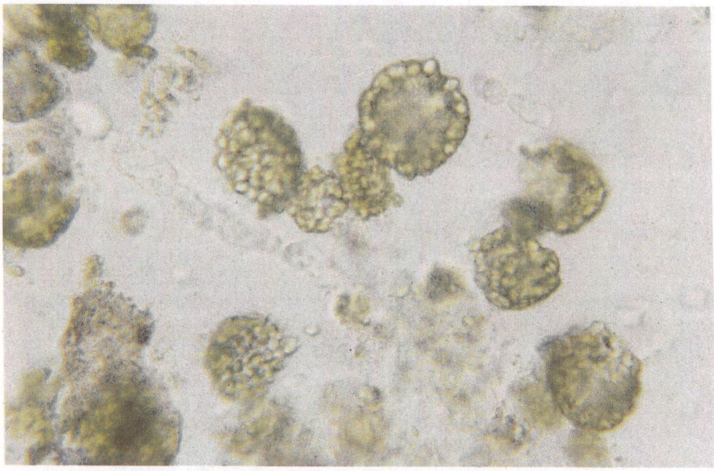

Plate 48: Protoplasts isolated from leaves of chilli

18.8. ISOLATION OF PROTOPLASTS FROM LEAVES OF *ANTHURIUM ANDREANUM* LIND. (Nirmala and Singh, 1996)

Plant Material

Leaf mesophyll tissue of axenic origin.

Preparation of the Plant Material

- Cut the fresh leaves from *in vitro* grown plantlets into small pieces of 1-2 mm^2.
- Place the leaf pieces in osmoticum for one and a half hours. (Osmoticum consists of 0.4M $CaCl_2.2H_2O$ with 0.3M mannitol in the presence of 0.5% MES buffer adjusted to a pH of 5.8. For efficient plasmolysis it is suggested to subject the leaves to vaccum infiltration in osmoticum for 2 minutes).

Enzymatic Isolation

- Treat the plasmolyzed material with 2.0% cellulase + 0.5% macerozyme dissolved in osmoticum. (10 ml of the enzyme solution is required to digest effectively 1.5 g of leaf tissue.)
- Incubate the mixture in dark for 12 hours.

Limitations

During protoplast isolation, it is usual to obtain broken protoplasts and also debris playing with changes and concentration of enzyme mixture, good protoplasts that is unbroken and round could be obtained.

A higher concentration of celluclast may be detrimental to the health (obtaining unbroken and round) of the protoplast.

Isolation of protoplasts is probably a good and exciting exercise for students. The culture of protoplasts and to obtain morphogenic responses from protoplasts is not an easy task except in classic examples such as tobacco and carrot.

19

PROTOPLAST FUSION AND CYBRIDIZATION

Plant protoplasts represent the finest single cell system that could offer exciting possibilities in the fields of somatic cell genetics and crop improvement. The lack of cell walls in protoplasts allows the plasma membrane of two or more protoplasts to come into intimate contact. Occasionally, spontaneous fusion of protoplasts may occur during their enzymatic isolation, but various techniques have been discovered to increase the frequency with which the fusion occurs. Protoplast fusion (somatic hybridization) offers a useful tool to the plant breeders to make crosses between sexually incompatible species for transfer of nuclear or cytoplasmic characters. The important aspect has been that incompatibility barrier does not exist during the cell fusion process at intraspecific, interspecific, intergeneric levels. Apart from overcoming the sexual barriers, it is useful in several Gramineae, Solanaceae and Cruciferae members to produce plants with resistance to diseases or tolerance to selective herbicides, to produce somatic hybrids with efficient organelle system and widen the genetic base.

19.1. CYBRIDIZATION

When two protoplasts fuse with their nuclei, a true hybrid is formed. However, often, the two nuclei survive independently in the mixed cytoplasms forming a heterokaryon. Chromosome loss can occur from one or both of the nuclei in the process. If the entire nucleus of one parent is lost, what left is one nucleus in a mixture of both cytoplasms. This product is called a **cytoplast** or **cybrid**. Thus, cybrids have nuclear genes of one parent and cytoplasmic genes of the other parent. Cybrids can be prepared by fusing a normal protoplast with an enucleated protoplast. This can be done by the inactivation of nucleus of one parent and subsequent fusion with normal protoplast. Presently, researchers are studying the use of cybrids to transfer resistance to some herbicides. This resistance character is carried on the chloroplast genome. Cytoplasmic organelles like chloroplasts and mitochondria possess DNA, which control the expression of male sterility. The transfer of mitochondrial genome can also control transfer of cytoplasmic male sterility. The study of cybrids is also important for the investigation of possible recombination within these extrachromosomal genomes.

19.2. PROTOPLAST FUSION

The fusion of protoplasts results in somatic hybrids and the process is known as somatic hybridization.

19.2.1. Spontaneous Fusion of Protoplasts

Occasionally, interspecific protoplasts have been found to fuse through their plasmodesmata during enzyme treatment. Protoplasts in a Petri dish have been observed to form large multinucleate masses. Protoplasts of young leaves often fuse spontaneously.

19.2.2. Induced Fusion of Protoplasts

This method can bring about the fusion of intra- and inter-specific protoplasts. For this, an inducing agent is necessary and these inducing agents are called fusogens. There are basically two means of induced fusion of protoplasts: by chemical induction and by electrofusion.

19.2.2.1. Chemical Fusion

The chemical fusion methods rely on the use of chemicals such as polyethylene glycol (PEG), high pH/Ca^{2+}, sodium nitrate, poly-L-ornithine, poly-D-lysine, glycerol, dimethyl sulphoxide. The most widely used fusogen in PEG. The effects of PEG are not specific and it promotes aggregation and fusion of protoplasts from the same or different species or intergeneric protoplasts. While chemical induced fusion often leads to successful fusion of the protoplasts, these chemicals cause a great deal of membrane disruption, and hence, the number of viable cells obtained are quite low after treatment.

19.2.2.2. Electrofusion

This is the most preferred method, as cell viability is not so adversely affected. The protoplasts are kept in between two electrodes and a potential difference is applied. Protoplasts fusion can be induced by applying extremely short, square wave electrical shocks. The major drawback here is that the technique requires expensive equipment.

Fig. 13 shows the diagrammatic representation of protoplast fusion.

19.3. SELECTION OF PROTOPLAST FUSION HYBRIDS

19.3.1. Visual Selection

This method can be used to select fusion products of protoplasts that have different distinct physical characters. For example, fusion between leaf mesophyll protoplasts of green colour and a cultured cell that has an anthocyanin containing red protoplast. A single fusion product between the two will have both chloroplasts and an anthocyanin vacuole together with the two nuclei. However, this method is very laborious and only a very limited number of fusion products can be selected.

19.3.2. Fluorescent Labels

In this method, fluorescent-labelled dyes are used to detect the fusion products. The two original protoplast cultures are pre-incubated in these dyes so that each group of protoplasts takes on a specific dye. After protoplast fusion, the products may be identified by their fluorescence characteristics under a fluorescence microscope.

19.3.3. Drug Sensitivity and Resistance

This method is employed where the two parental species are sensitive to a drug while the fusion product is resistant.

19.4. PROCEDURE FOR CHEMICAL INDUCED FUSION OF PROTOPLASTS

1. Place a drop of each protoplast suspension in a sterile Petri dish, gently shake the plate to ensure proper mixing of suspensions from callus or leaf protoplasts. The concentration of each species should be about 1×10^5/ml before mixing.
2. Wait for 5 minutes to allow the protoplasts to settle to the bottom of the Petri dish.
3. Slowly add 300-450ml PEG solution to the edge of the protoplast suspension, then in the centre of the protoplast and wait for 15 minutes.
4. Add drop-by-drop 1 ml of 0.7M mannitol solution to dilute the PEG solution.
5. Raise one side of the plate and wash the protoplast clumps adhering to the plastic surface of plate with 9 ml of 0.7M mannitol.
6. Remove the PEG and mannitol residues from the Petri dish. Add a few drops of mannitol solution to the fused cells.
7. After 5 minutes, observe the fused product on an inverted microscope. The process of adhesion or agglutination of protoplasts in various combinations can be seen.

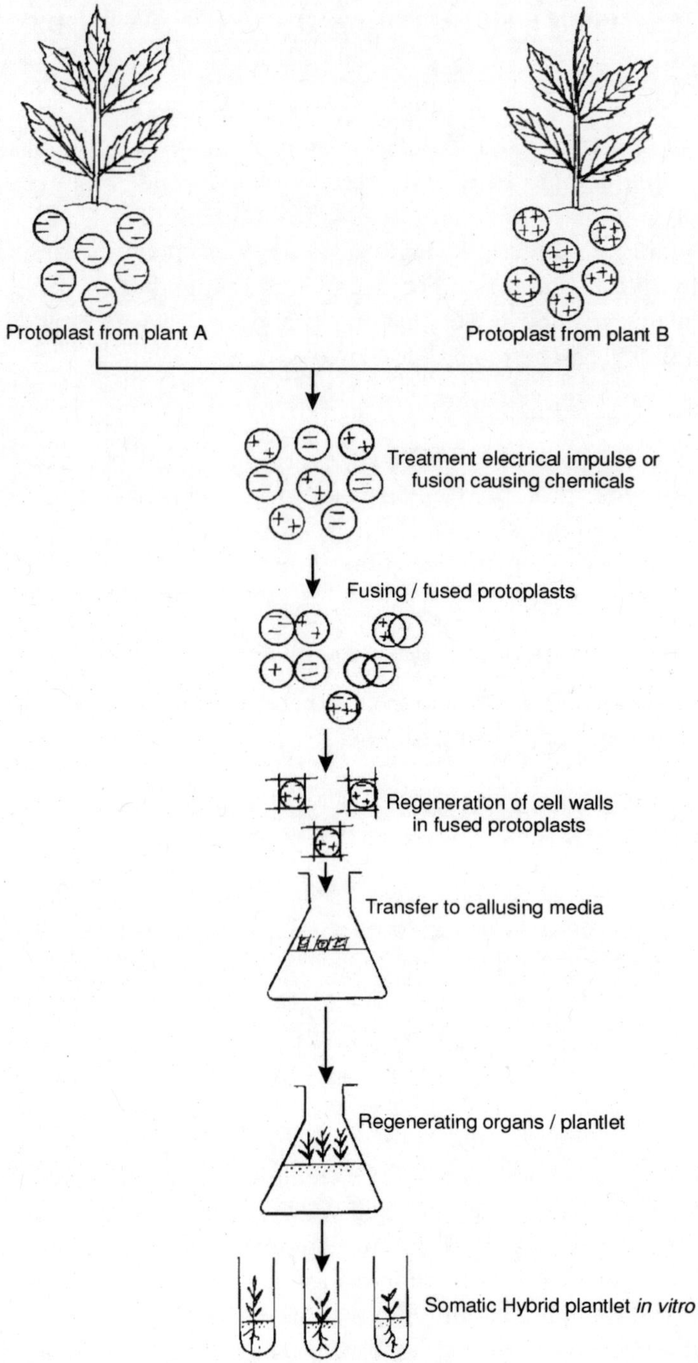

Fig. 13: Diagramatic illustrations of protoplast fusion and regeneration of stomatic hybrid plants.

8. Culture the protoplasts in a suitable medium and incubate in dim light.
9. After 2 to 3 weeks, regenerate shoots and roots and transfer to *ex vitro* conditions.

Observations to be recorded

1. Observe the fused and unfused protoplasts for the synthesis of cell wall and first cell division.
2. Calculate the frequency of homokaryons/heterokaryons.

20

Storage Organ Culture

Many ornamental and crop species are normally propagated, stored and planted in the form of vegetative storage organs. Such organs, which includes bulbils, cormlets, protocorms and miniature tubers, when produced *in vitro*, often provide a convenient means of micropropagation and/or genotype storage. Methods of obtaining storage organs vary according to the kind of tissue being cultured. Some storage organs formed *in vitro* can be planted *ex vitro* directly into the soil.

20.1. THE PRODUCTION OF BULBILS AND CORMLETS

Species that naturally produce bulbs, can be induced to form small bulbs (bulbils or bulblets) in culture. They arise on cultured tissues either at the base of a previously formed vegetative shoot or as a directly initiated storage organ with no extended vegetative leaves. Buds giving rise to bulbils may arise adventitiously on pieces of leaf, on inflorescence stalks, or on ovaries, but particularly on detached pieces of bulb scale. Strong dominance of the main shoot apex often prevents the formation of axillary buds at the bases of bulb scales, but buds capable of giving rise to bulbils are freely produced in axillary positions when bulb scale pieces are cultured.

Instead of producing storage organs composed of swollen leaf bases (bulbs), some monocotyledons store food reserves in swollen stem bases (corms). Small corms (cormlets) may be formed directly on explanted tissue or on callus in culture, or they may be produced on rooted plantlets grown in culture jars until the leaves senesce. Cormlets formed *in vitro* can be planted in the soil or used to start new *in vitro* cultures.

20.2. PROTOCORM FORMATION BY ORCHIDS

Upon seed germination, the orchid embryo swells to form a small storage organ called protocorm, possessing an apical meristem and a leaf primodium. The protocorm becomes green and accumulates carbohydrate reserves that are used in the growth of orchid seedlings, which are ultimately formed. Structures resembling protocorms were found to be formed on shoot tips cultured on simple media without growth regulators, sometimes in great numbers. These protocorms could be divided and cultured and the regenerated ones could develop into plantlets. In some orchid genera, protocorms are not only formed on apical meristems but also directly form on explants such as leaves.

20.3. MINIATURE TUBERS

Many small dormant tubers can be obtained from virus-free shoots cultured *in vitro*. These miniature tubers can readily be removed from culture flasks, stored without much aseptic precautions, and then distributed to growers for the production of plantlets. Potato tubers has been found to arise on cultured shoots *in vitro*, under the influence of high cytokinin levels.

20.4. *IN VITRO* TUBERIZATION IN POTATO

Potato (*Solanum tuberosum*) is a crop of considerable importance, ranking fourth in the world production after wheat, rice and maize. The traditional method of propagation of potato is by using seed tubers. Because of the vegetative method of perpetuation of tuber generations, pathogens are easily transmitted. The conventional method of potato propagation is also constrained by high cost associated with the purchase, storage and transportation of seed tubers. Advances in plant tissue culture and micropropagation techniques have a definite and indispensable advantage over the conventional method of propagation by the induction of *in vitro* microtubers.

20.5. ADVANTAGES OF USING *IN VITRO* MICROTUBERS

1. Short time period to produce the seed tubers.
2. Reduced number of field generation required resulting in a better quality seed.
3. Ease of handling, storing and planting.
4. Renders maximum uniformity of clones.

20.6. PROTOCOL FOR *IN VITRO* MICROTUBER PRODUCTION IN POTATO (Kedisso and Sathyanarayana, 1998)

Explant Preparation

- Sprouts, to be used as explants can be collected from one year old microtubers.
- Wash the sprouts with doubled distilled water and one drop savlon for 10 minutes.

- Rinse twice in double distilled water and dip in 70% alcohol.
- Rinse again thoroughly in doubled distilled water.
- Inside the laminar flow chamber, wash the sprouts in sterile water once followed by washing in 0.1% mercuric chloride for a minute. Rinse the material in sterile water thrice to remove the traces of sterilent.

Initiation of Cultures

- Inoculate the sprouts in full strength MS medium supplemented with 0.25 mg/L GA_3, 0.03 mg/L NAA and 2.0 mg/L Calcium pantothenate and 3% sucrose.
- Keep the cultures under 16h photoperiod and a temperature of 26 ± 3°C.
- Subculture every four weeks.

In vitro Tuberization

- Excise nodal sections from four-week old *in vitro* derived plantlets of six nodes each, after removing the apical and basal parts.
- Transfer them to the tuberization media (MS media with 10 mg/L BAP, 8% sucrose and 2 g/L activated charcoal).

Plate 49: Microtubers produced *in vitro*
(After Kedisso and Sathyanarayana, 1998)

- Incubate the cultures in a growth chamber with a temperature of 26 ± 3°C and 16 hours photoperiod for a month after which they are transferred to another growth chamber with 8 hours photoperiod and 15/15°C day/night temperature and light intensity of 5000lux for 8 weeks. (Plates 49 and 50)

Plate 50: *In vitro* microtubers of potato
(After Kedisso and Sathyanarayana, 1998)

21
In Vitro Germplasm Conservation

Ever since primitive man learned the art of farming and realized the economic utility of plants, he started saving selected seeds or vegetative propagules from one season to the next. In fact, he was practicing a type of germplasm conservation and management. The primitive cultivars and their wild relatives constitute a pool of genetic diversity, which is invaluable for breeding programmes. However, the population pressure and needs of man have steadily increased over time leading to persistent exploitation of natural resources. The pressure thus created led the way for gradual exclusion of traditionally used genotypes. In this way, some of the valuable gene pools might be lost unless coordinated efforts are made towards the conservation of genetic stocks all over the world. Realizing the danger of erosion of genetic resources, the objective of providing necessary support to collection, conservation and utilization of the plant genetic resources is a major concern of the world today.

Though the interest in the conservation of germplasm was triggered as early as 1930, it was limited to certain individual scientists only. It was only after 1964, that the term "germplasm" or plant genetic resource was coined. The conservation of germplasm work took a definite path with the establishment of International Biological Programme at the United Nations Organization, which coordinated its activities with the Food and Agricultural Organization.

Diverse germplasm conservation methods are pursued today, which can be broadly divided into *in situ* and *ex situ* methods. *In situ* conservation involves conservation of the germplasm in their natural habitats in the wilderness, reserves, protected areas, and within traditional farming systems. This method is especially useful in conservation of species that are nearing extinction. However, this method is difficult to adopt if the species is distributed over a large geographical area. Further, there is the risk of genetic wipe-out.

Ex situ conservation involves removing the plant genetic resources from their natural habitats and placing them under artificial storage conditions. *In vitro* germplasm conservation has been widely adopted as an important means of *ex situ* conservation. *In vitro* method of conservation, though not a panacea for all the problems encountered in the conventional methods, does offer considerable advantages.

21.1. ADVANTAGES OF *IN VITRO* GERMPLASM CONSERVATION

- The genetic stability of the cultures will be higher than those maintained by the conventional methods.
- Space and maintenance required is less than the conventional method.
- The plants maintained are ideal for international exchange and distribution of germplasm since they will be acceptable to the plant quarantine authorities.
- It has utility in the conservation of species having recalcitrant seeds or those propagated vegetatively.
- The diseases and pests can be successfully eliminated, even the viruses can be eliminated using thermo-or chemo-therapy.
- Rapid clonal multiplication of the conserved material is possible.
- Labour and cost savings

21.2. *IN VITRO* GERMPLASM EXCHANGE

In vitro techniques are in use for several years for the international distribution of plant genetic resources. The techniques used are based on standard mass propagation procedures, with minor but the following important modifications to increase the structural stability during transit.

(*a*) The concentration of agar or other gelling agent used for preparing the culture medium may be raised to increase its firmness.

(*b*) The plantlets may be transferred in sterile heat-stable polyethylene bags rather than fragile glass or plastic containers.

21.3. THEORETICAL BASIS OF *IN VITRO* GERMPLASM CONSERVATION

During the long term storage of living material, changes in chemical composition or physical structure may lead to cell death or permit accumulation of mutations. However, only reversible processes can be tolerated. Biochemical reactions in the cells require water as substrate otherwise the processes are stopped. Water can be eliminated by transferring the liquid water to the solid or gaseous state or by substituting water with other substances.

21.4. TRANSFER OF WATER TO THE SOLID STATE

Freezing of water occurs at wide range of temperatures. While pure water becomes ice at 0°C, the cell water needs much lower temperatures because of freezing point depression by salts or organic molecules. The type of crystal formation during freezing process and during storage controls the survival of the tissue.

21.5. TRANSFER OF WATER TO GASEOUS STATE

The transfer of water to the gaseous state results in dehydration. In most cases, this leads to cell death. However, in some specially adapted organisms or tissues, dehydration occurs naturally.

21.6. THE SUBSTITUTION OF WATER

Sometimes, water can be substituted by a lot of organic solvents in cells. Little is known about the biochemical background of this conservation method.

21.7. PLANT MATERIAL USED FOR *IN VITRO* CONSERVATION

Almost all types of culture, viz., callus, suspension cultures, pollen, meristem, shoot tip and somatic embryos can be conserved by *in vitro* methods. However, callus and cell suspensions are rarely used since they are more prone to somaclonal variation. Meristems and shoot tips are preferred as plant material as they are found to retain the genetic stability even after considerable storage period.

The type of plant material used depends on the *in vitro* method of conservation being used. Generally, smaller the size and less complex the structure, better is the chance of survival with a low risk of somaclonal variation.

21.8. *IN VITRO* PRESERVATION OF SEED PROPAGATED SPECIES

In vitro technologies are becoming important in maintaining crops with descication-sensitive seeds. Embryo culture may be used in the preservation of such desiccation sensitive seeds as isolated embryonic axes from mature seeds can survive partial desiccation and cooling to low temperatures.

21.9. *IN VITRO* PRESERVATION OF VEGETATIVELY PROPAGATED SPECIES

The use of tissue culture in germplasm preservation is usually applied to vegetatively propagated crops. There are basically three main approaches to *in vitro* germplasm conservation:

21.9.1. Maintenance of Cultures at Normal Growth Conditions

Plants, which respond to *in vitro* propagation, can be maintained indefinitely with regular subculturing. Since this method is quite tedious, this is not a preferred method of conservation. However, in certain crops like coffee, the *in vitro* propagated plants have shown very slow growth and thus can be used for conservation.

21.9.2. Maintenance of Cultures at Minimal Growth Conditions

This method is more applicable to *in vitro* germplasm conservation. The growing conditions are altered such that the plant material continues to grow at a very slow pace thereby increasing the duration between two subcultures. This can be achieved by:

(*i*) Reducing the nutrient level, especially the carbon source in the media.
(*ii*) Reducing the storage temperature.
(*iii*) Including sub-lethal level of growth retardants.
(*iv*) Inducing osmotic stress in the culture media using sugar/mannitol.

Advantages

- The duration between the subcultures is increased, so less maintenance is required.
- The risk of contamination is reduced.
- The material maintained can always be taken out for further multiplication purpose.

Disadvantages

- The conditions required to induce slow growth need to be standardized for every species separately.
- In most of the species tried, the interaction of the various factors like temperature, growth retardants etc., have not been fully evaluated.
- Though there is reasonable information available on the duration of subculturing intervals, we do not know how many subcultures can be subsequently applied and whether the intervals remain constant over time.
- There is knowledge gap in regard to the genetic and phenotypic stability of the material exposed to slow growth conditions overextended period of time. This is especially true in case of unorganized cultures.

21.9.3. Cryopreservation

Storage of plant material under very low temperatures using liquid nitrogen (-196°C) and/or liquid nitrogen vapour (-150°C) is called cryopreservation.

21.9.3.1. In vitro Preservation of Pollen

The major use of germplasm is for improvement of plant by sexual processes, and thus requires controlled crosses between desired individuals. Supplying pollen to the user facilitates this process, as the user does not have to grow the plant and wait for the flowering of the male parent to occur. Pollen from many species is desiccation tolerant and exhibits storage characteristics similar to desiccation tolerant seeds. In the case of pollen from species that are desiccation sensitive, cryogenic storage is feasible but only within a narrow range of moisture contents. Pollen preservation is a supplement to seed or clone storage, not a substitute for them.

21.9.3.2. Preservation of Somatic Embryos

The technique of somatic embryogenesis has been developed for elite genotypes of numerous crop species leading to the production of a large number of synchronously developing embryos. These embryos can be encapsulated in a bead of gel (e.g., Calcium alginate) containing nutrients and fungicides thereby forming synthetic seeds which, theoretically, can be stored and sown directly *in vivo* similar to true seeds. The application of synthetic seed technology to somatic embryos or shoot tips appears to be of interest in a germplasm conservation context. However, only a limited number of short term to medium term storage experiments have been performed with encapsulated material. Encapsulated axillary buds of mulberry (*Morus indica* L.) and somatic embryos of sandal wood (*Santalum album*) can be stored for 45 days at 4°C. Storage for longer periods was achieved if beads were placed in liquid medium at low temperature. Under these conditions, mulberry apices could be conserved for 80 days (Shigeta *et al.*, 1993).

21.10. MOLECULAR TECHNIQUES OF GERMPLASM CONSERVATION

Molecular technologies based on DNA extraction and storage offer new ways of germplasm conservation. In the future, it may be possible to envisage the storage of specific gene sequences within a broad complimentary conservation strategy.

Cryopreservation

Cryopreservation is the storage at ultra low temperatures in a cryogenic medium such as liquid nitrogen. It provides an indefinite storage system for organs, tissue or cells at ultra low temperature (at -196°C using liquid nitrogen) by reducing metabolism significantly with no biological deterioration. Rather than the time storage being the important factor, the real critical stages are the transition to and from the frozen state. This is a highly unnatural transition and in case of most plant materials, measures must be taken to protect both cellular and tissue structures against the deleterious effects. These measures need to be applied at each of the stages of cryopreservation.

21.11. STAGES OF CRYOPRESERVATION (Withers, 1991)

Plant material
↓
Pre-growth stage
↓
Cryoprotection
↓
Cooling
↓
Storage
↓
Thawing
↓
Recovery

21.11.1. Pre-growth Stage/Pre-treatment Stage

This is a period when hardiness can be induced in the plant species to increase its freeze tolerance. Shoot tips isolated from plants that have been cold acclimated survive the cryogenic protocol in higher percentages than those not acclimated. This is useful only in those species that have the genetic capacity to cold acclimate. Survival of excised shoot tips can also be enhanced by exposure to elevated sucrose levels. These treatments might alter the intracellular contents of critical metabolites (particularly sugars), cell water content, and physiological characteristics of the cell like membrane permeability. Sometimes, such treatments help in the simplification of the procedures to follow. In case of somatic embryos of oil palm, using high levels of sucrose (0.25M) eliminated the need for cryoprotectants and reduced the dependence on precisely timed cooling rates.

21.11.2. Cryoprotection

After the pre-treatment stage, shoot tips are exposed to cryoprotectants. The concentration of cryoprotectant used and the time and temperature of exposure vary with cooling strategy employed. There are two types of compounds used as cryoprotectants:

(*a*) Compounds, which penetrates the cell rapidly, e.g., Dimethyl sulphoxide

(*b*) Compounds, which cannot penetrate the cell or which penetrates the cell very slowly, e.g., Low molecular weight substances like sucrose and high molecular weight substances like polyvinyl pyrrolidone.

It has been postulated that both types of compounds might work in the same way by protecting the surface membranes. Cryoprotectants facilitate the flow of water across the cell membrane, and protect both molecular and gross structures through a range of modes of action, including colligative effects and free radical scavenging.

21.11.3. Cooling and Storage

Two cooling regimes are mainly employed in cryopreservation:

21.11.3.1. Two-step/Slow/Optimum Rate Cooling

This method involves cooling of the cryoprotected shoot tips at about 0.25 to 1.0°C/min to about -35°C prior to immersion in liquid nitrogen. For this method, shoot tips are incubated in either 1 to 2M concentration of a suitable permeating cryoprotectant such as dimethyl sulphoxide or ethylene glycol.

21.11.3.2. Rapid Cooling/Vitrification

Here, shoot tips are exposed to much more concentrated solutions of cryoprotectants and then are immersed directly from about 22 or 0°C into – 196°C. The term vitrification refers to transition of water directly from the liquid phase into an amorphous phase, while avoiding the formation of crystalline ice during cooling. The advantages of vitrification are that it is a simple technique and does not require an expensive apparatus, it may avoid the damaging consequences of ice formation, and it may be applicable to larger pieces of tissue than are used for two-step cooling method. Vitrification is more appropriate for complex organs (shoot tips or embryos) which contain a variety of cell types, each with unique requirements under conditions of freeze-induced dehydration.

A major variation of vitrification uses an **encapsulation/dehydration sequence**. This technique is based on the technology developed for the production of synthetic seeds. The plant material is encapsulated in an alginate gel, cultured in sucrose solutions, dehydrated in an airflow and then usually rapidly cooled.

21.11.3.3. Thawing

Thawing by rapid warming from -196°C to about 22°C is generally beneficial for obtaining high levels of survival for samples cooled by either two-step cooling or vitrification. Rapid warming minimizes any destructive ice crystal growth.

21.11.3.4. Recovery of Cultures

After thawing, the plant material is washed in sterile media several times to remove the cryoprotectant adhering to the surface and is then transferred to the culture medium for plantlet regeneration. Treated plant material is preferably cultured in the dark or dim light for a few days to minimize any oxidative stress that might occur due to damaged mitochondria or chloroplasts.

Table 9 gives examples of few horticultural crops subjected to cryopreservation.

Table 9: Horticultural crops subjected to cryopreservation

Species	*Culture System*	*Pre-growth*	*Cryoprotection*	*Cooling, Storage, Thawing*	*Recovery*
Solanum tuberosum	Shoot tip	Shoot tip dissected from glasshouse or *in vitro* plants; incubate overnight in standard liquid medium	10% DMSO	–0.20°C/min to –35°C; transfer to liquid nitrogen; thaw in water bath at +37°C	Wash twice with liquid medium; transfer to semi-solid medium
Pyrus sp.	Shoot tip	Dissect shoot tips from in vitro plants; pre-grow for 48h in medium containing 5% DMSO	10% polyethylene glycol +10% glucose + 10% DMSO	–1°C/min to –40°C; transfer to liquid nitrogen; thaw in water bath at +40°C for 1 min; transfer to +23°C	Wash twice with liquid medium; transfer to semi-solid medium
Dianthus caryophyllus	Shoot tip	Dissect shoot tip from cold hardened plant	5% DMSO	Place in ampule and plunge into liquid nitrogen	Culture under standard conditions

22
In Vitro Production of Plant Secondary Metabolites

The potential use of plant tissue culture for the production of compounds valuable to the industry has long been a subject of interest for biotechnologists. Considering the high economic and pharmacological importance of the secondary metabolites, industries are deeply interested in utilising plant tissue culture technology for the large-scale production of these substances. Cultured plant cells have been known to produce biochemicals of interest since 1950s, but the pioneering attempts achieved little success. Later refined culture systems of cell suspensions and calli, and the successful selection of cells producing high amounts of secondary compounds have propelled tissue culture production of secondary metabolites into the realm of economic feasibility. Shikonin, a napthoquinone from *Lithospermum erythrorhizon* was the first commercial product to be obtained from cell cultures.

In some cases, where the metabolite is only produced in specialized plant tissues or glands, production has not been successfully stimulated in cell culture. An alternative approach to the production of secondary metabolites *in vitro* in such cases is the use of organ culture. For instance, in ginseng, saponins and other valuable metabolites are specifically produced in the roots and hence, *in vitro* root culture is required.

The three long-standing classical examples of commercially viable production of a secondary metabolite *in vitro* are ginseng, saponins, shikonin, and berberine. Ginseng is produced in large-scale root cultures, whereas, the other two products are produced in highly coloured cell cultures. Tremendous research and development efforts have advanced a number of other *in vitro* derived secondary metabolites to semi-commercial status, including vanillin and taxol production in cell cultures. Engineers and biologists are currently joining forces on a global scale to develop new strategies for streamlining the critical bioprocesses.

Advantages of *in vitro* Secondary Metabolite Production

- There are several distinct advantages for producing valuable secondary metabolites *in vitro*, rather than *in vivo* in the plant.
- Plant cells are relatively easy to grow and can be kept under strict controlled nutritional and environmental conditions. Hence, the uncertainties of climate and soil can be avoided.
- Secondary products *in vitro* can be generated on a continuous year round basis.
- Production is reliable and predictable (independent of ambient weather) and at times, may exceed that found in nature.
- Plant cell lines can be recurrently selected to amplify the productivity of the cell culture.
- Suspension cultures offer a very effective way of incorporating precursors, which are often difficult to administer to the plant growing in nature.
- Disagreeable odours or flavours associated with the crop plant can be modified or eliminated *in vitro*.
- Extraction from the *in vitro* tissues is much simpler than extraction from organized, complex tissues of a plant.
- Offers the rare opportunity to tailor the chemical profile of a phytochemical product by manipulation of the chemical or physical microenvironment.

22.1. REQUIREMENTS FOR SUCCESSFUL *IN VITRO* SECONDARY METABOLITE PRODUCTION

- The rate of cell growth and biosynthesis should be high enough to give a good production of the final product in a short period of time.
- The cultured cells should be genetically stable to give a constant yield of the product.
- The metabolites should be accumulated in cells without being catabolized rapidly or preferably, they should be released into the medium.
- Production cost including the culture medium, precursor and chemical extraction should be low enough to be profitable to the industries.

22.2. PLANT CELL CULTURES

Cultured plant cells often produce reduced quantities and different profiles of secondary metabolites when compared with the intact plant and these quantitative and qualitative features may change with time. The poor product expression is often attributed to a lack of differentiation in cultures. On the other hand, there are cases of cultures that over-produce metabolites compared with the whole plant (Table 10).

There are a number of examples of cultured cells producing metabolites not observed in the plant, e.g., *Lithospermum erythrorhizon* cultures have been observed to synthesize rosmarinic acid. It has become apparent that the choice of original plant material having high yields of the desired phytochemical may be important in establishing high-yielding cultures. Furthermore, the need to repeatedly screen for high-producing lines (due to inherent instability of cell lines) has been emphasized, although the nutritional composition of the medium is also important. Thus, a variety of approaches are being investigated by many researchers to increase productivity of useful plant metabolites in plant cell cultures.

Table 10: Secondary Metabolites Produced in High Levels by Plant Cell Cultures

Compound	*Plant species*	*Yields (% Dry wt)*		*Culture*
		Culture	*Plant*	*type*
Shikonin	*Lithospermum erythrorhizon*	20	1.5	Suspension
Ginsenoside	*Panax ginseng*	27	4.5	Callus
Anthraquinones	*Morinda citrifolia*	18	0.3	Suspension
Ajmalicine	*Catharanthus roseus*	1.0	0.3	Suspension
Rosmarinic acid	*Coleus blumeii*	15	3	Suspension
Ubiquinone-10	*Nicotiana tabacum*	0.036	0.003	Suspension
Diosgenin	*Dioscorea deltoides*	2	2	Suspension
Benzylisoquinoline Alkaloids	*Coptis japonica*	11	5-10	Suspension
Berberine	*Thalictrum minor*	10	0.01	Suspension
Berberine	*Coptis japonica*	10	2-4	Suspension
Anthraquinones	*Galium verum*	5.4	1.2	Suspension
Anthraquinones	*Galium aparine*	3.8	0.2	Suspension
Nicotine	*Nicotiana tabacum*	3.4	2.0	Callus
Bisoclaurine	*Stephania cepharantha*	2.3	0.8	Suspension
Tripdiolide	*Tripteryqium wilfordii*	0.05	0.001	Suspension

22.2.1. Immobilized Cell Culture

Instead of being grown in suspension cultures, cells can be entrapped or immobilized within a matrix, or upon an inert supporting material. This can have a number of advantages like:

- Cells are protected from the shearing forces normally found in bioreactors and can be grown in high densities.
- Usually, immobilized cells are able to synthesize chemicals over longer periods than in suspensions so that continuous, rather than batch production becomes a possibility.
- Immobilization also keeps cells away from the interface of the medium with air, where, coagulation, frothing, and sometimes deactivation can occur.

22.3. CULTURE OF DIFFERENTIATED ORGANS

Culture of roots or shoots are capable of synthesizing compounds, which may not be found in undifferentiated cells, and also in larger quantities. Root cultures have been found to be highly suitable for the production of certain chemicals.

22.3.1. Hairy Root Culture

A fairly recent development in plant organ culture, consisting of culture of highly branched roots of a plant. A plant tissue is treated with a culture of the bacterium, *Agrobacterium rhizogenes*, which transfers part of its own plasmid DNA to the cells of an infected plant. This alters the plant's metabolism, including alterations in hormone levels, which in turn cause the explant to grow highly branched roots from the sites of infection. The roots branch much more frequently than the usual root system of that plant, and are also covered with a mass of tiny root hairs. Their most significant feature is that they produce secondary metabolites at levels similar to those made in the original plant. Thus they can be used as replacement plants for making such compounds as food flavours or fragrances. The main application of this technique is the over-production of secondary metabolites and biotransformation of chemicals. Hairy root cultures have been shown to produce compounds that have not been identified in non-transformed roots.

22.3.1.1. Characteristics of Hairy Root Cultures

- Stable metabolite production over successive generations
- Inherent genetic stability reflected in stable productivity
- Plagiotropism of hairy roots is one of the readily observed characteristics
- An exponential increase in biomass occurs as a result of lateral root formation and a consequent exponential increase in the number of elongating tips
- Possibility of genetic manipulation to increase biosynthetic capacity

These characteristics of hairy root cultures have initiated considerable interest both as fundamental research tool and as a source of valuable products.

22.3.1.2. Secondary Metabolite Production by Hairy Roots

The diversity of secondary metabolites found in roots represent an immense biochemical potential, most of which is still waiting for thorough exploration. With the availability of a system, which involves the culture of roots obtained by transformation with *Agrobacterium rhizogenes*, there is renewed interest for *in vitro* production of fine chemicals. The primary advantage of root cultures is that they reproduce the tremendous biosynthetic capability of roots found in the intact plants. These hairy roots have the same metabolic features as normal root cultures, yet are as rapid growing as cell suspension cultures. These roots grow much faster than normal roots. Hairy roots can be cultured in a simple defined medium without addition of growth hormones and can grow from low inocula to high final biomass densities with only minimal lag phase.

22.3.1.3. Scaling-up of Hairy Roots

One of the important limitations of commercial exploitation of this technique is the development of technologies for the scale-up of these root cultures. Bioreactors used for root cultures are basically the same as that used for plant cell suspensions. There must be sufficient mixing and mass transfer to provide adequate nutrient availability under operating conditions, which are not detrimental to growth. Unlike cell suspensions, where cells are suspended in liquid medium, roots are either fixed on a support upon inoculation or are stationary by virtue of the entwined matrix formed during growth.

22.4. BIOTRANSFORMATION

Biotransformation is the process whereby a substance is changed from one chemical to another (*transformed*) by a chemical reaction within the tissue. The terms **Metabolic transformation** and **Bioconversion** are frequently used for the biotransformation process. Cultures can be fed with the chemicals which bioconvert them into more valuable products, which are then of use as pharmaceuticals or fine chemicals. The α-hydroxylation of digitoxin by *Digitalis lanata* is probably the most well known bioconversion as it has been used to obtain a commercially valuable product.

22.5. STIMULATION OF CHEMICAL PRODUCTION

22.5.1. Elicitation

In general, the biochemical yields from cultured plant cells can be improved by two means:

1. By increasing cell biomass, yield per unit volume of culture. However in most cases, the culture conditions favouring growth suppress biochemical production.
2. By increasing the biochemical content of the cell biomass.

Elicitation is one of the important strategies towards increasing the biochemical content of the cells and thereby increasing the secondary metabolite yield.

Some molecules stimulate the production of secondary metabolites; these are called **elicitors** and the pheonmenon is known as **elicitation**. Low concentrations (50-250 mg/L) of elicitors added to the culture medium dramatically enhance secondary metabolite production by plant cells. Many biotic (pectin, pectinic acids, cellulose—all endogenous elicitors, chitin, chitosan or glucans—all exogenous elicitors) and abiotic substances (UV irradiation, salts of heavy metals, other chemicals) act as elicitors and trigger the production of secondary metabolites in plants.

The elicitation process is accompanied by an increase in biosynthetic enzyme activity, which is regulated at the level of transcription. Activation of gene transcription requires recognition of the elicitor at the plant/cell surface by a cell receptor and subsequent transduction of the signal to the nucleus; probably mediated by second messengers.

22.5.2. Precursor Feeding

Product yield may be increased by feeding cells with a precursor of a series of biosynthetic reactions. However, the compound supplied must usually be a fairly immediate precursor of the final product, otherwise it will tend to be utilized in alternative synthetic pathways.

23
Acclimatization of Micropropagated Plantlets

Substantial numbers of micropropagated plants do not survive transfer from *in vitro* conditions to field environments due to lower relative humidities, higher light levels, and septic environments that are stressful to micropropagated plants compared to *in vitro* conditions. Under *in vitro* conditions, plantlets or shoots are not subjected to physiological stress since a carbon source is provided, which reduces the need for photosynthesis. The aseptic environment *in vitro* also reduces the stress of pathogenic organisms. These conditions combine to produce plants, which are unsuited for survival if planted directly in the field. Hence, these plantlets are to be acclimatized before they can survive the harsh conditions *in vivo*.

The terms acclimatization and acclimation though similar, cannot be used as synonyms. The word acclimation denotes the process during which the plants or other organisms become adjusted or accustomed to a new climate or situation as a result of natural processes. Plants growing outdoors, for example, undergo acclimation to cold when winter approaches. Acclimatization, on the other hand, implies that humans have interceded and guided the process of adjustment. Since propagators are inherently involved in weaning plants from *in vitro* conditions for establishment in the field, the process is termed acclimatization.

23.1. WATER STRESS

Water stress can result from excessive transpiration from above ground parts of plants, especially the leaves, or from inadequate uptake of water from the roots. The cuticle and stomata are the primary routes of water loss from the leaves.

23.2. THE CUTICLE

The cuticle is a membrane composed of a cutin matrix together with embedded and surface waxes that covers above ground tissues of plants. The primary function of the cuticle is to limit transpirational loss of water. Water permeability through the cuticle is influenced primarily by the structure and amount of cuticular and epicuticular waxes.

Studies have indicated that the crystalline structure of epicuticular waxes on leaves of micropropagated plants was significantly different from that of greenhouse grown plants. The lack of crystalline structure of wax was also correlated to less amount of epicuticular wax, compared to greenhouse grown plants. The lack of epicuticular wax on plants *in vitro* is due to the high humidity within the culture vessel. The reduced amount of epicuticular wax was directly correlated to substantially increased water loss in cultured shoots.

23.3. STOMATA

Stomatal structure and functioning have been implicated in the water imbalance exhibited by micropropagated plants upon removal from culture. Stomata present on *in vitro* plantlets are raised, rounded guard cells compared to normal elliptical, sunken guard cells in the field grown plants. The stomata was also found to be wide open in the *in vitro* conditions (Plate 51). (Sathyanarayana, 1993).

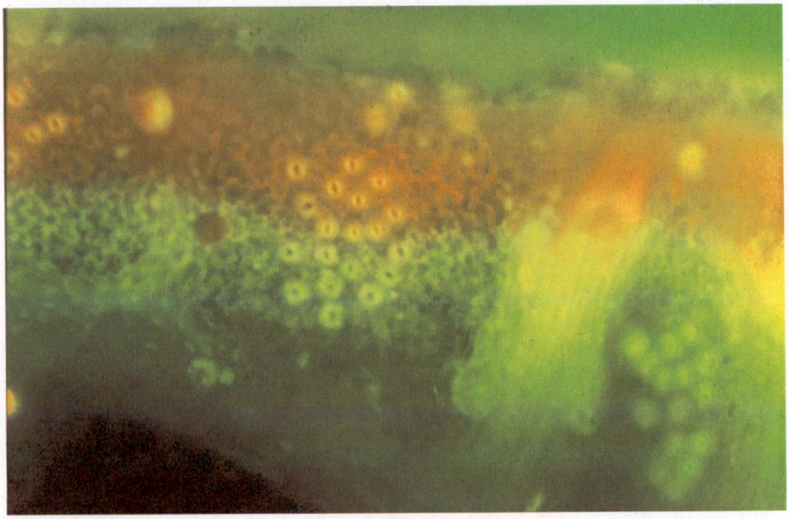

Plate 51: Open stomata found in leaf surface of *in vitro* derived jackfruit plantlets

Many researchers have reported that stomata of cultured plants show a characteristic inability to close when first removed from culture. It has been postulated that in spite of the fact that most stomata closed within 12 to 24h after removal of shoots from culture, the amount of water lost over 24h was equivalent to two to three times the initial weight of the plants (Shackel, *et al.*, 1990).

23.4. HYDATHODES

In some micropropagated plants like blackberry and strawberry, the hydathodes appeared to be open and exhibited guttation contributing to adaxial water loss.

23.5. LEAF, STEM AND ROOT ANATOMY OF MICROPROPAGATED PLANTLETS

Histological studies have revealed that leaves of *in vitro* developed plantlets have a characteristically poorly developed palisade layer with a significant amount of mesophyll air space (Plate 52) compared to greenhouse grown plants (Sathyanarayana, 1993). The chloroplasts of *in vitro* grown plants appear flattened with disorganized grana lacking starch granules.

The development of new leaves after removal from culture is considered to be important for the vigorous growth of the plants. During acclimatization, leaves present as primodial leaves *in vitro* assumes intermediate characteristics

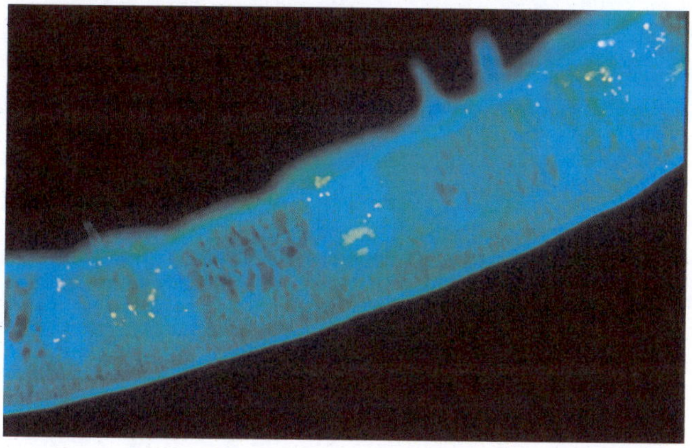

Plate 52: Section of leaf showing palisade tissue

between leaves of plants grown *in vitro* and greenhouse or field grown ones (Grout and Aston, 1978).

Stems of plants under *in vitro* conditions are found to be more slender with considerably less collenchyma and sclerenchyma supportive tissue than plants that are grown in the field. Roots produced by these plantlets *in vitro* were slender, covered with root hairs and have much less periderm (Donnelly *et al.*, 1985). The connection between roots and shoots of such plantlets are incomplete resulting in inefficient water transfer between the roots and shoots.

23.6. PHOTOSYNTHESIS

Plantlets undergo a rapid and extreme change in physiological functioning when they are removed from *in vitro* culture to soil mixtures. Thus, the plantlets are forced to change from being heterotrophic to autotrophic, and photosynthesis becomes paramount for survival.

Grout (1988) postulated that cultured plantlets could be divided into two discrete groups based on photosynthetic response. In one group, represented by strawberry and cauliflower, leaves formed *in vitro* never developed photosynthetic capability if they grew while the plant was on medium containing sucrose. After transfer to medium lacking sucrose, these leaves functioned as storage organs supplying carbon, stored as starch, to the new developing leaves. The persistent leaves were not photosynthetically competent and eventually died. The second group included species such as *Dieffenbachia* and had persistent leaves that eventually adapted to autotrophic conditions, with concomitantly increasing rates of photosynthesis.

The generally low rates of photosynthesis in plantlets grown *in vitro* have been attributed to low RuBP activity (Grout, 1988), low light, and inadequate gas exchange. The low RuBP activity was presumed to be due to the presence of sucrose during the development of leaves.

23.7. TECHNIQUES OF ACCLIMATIZATION

The physiological and anatomical characteristics of micropropagated plantlets necessitate that they are gradually acclimatized to the environment of the greenhouse or the field. Techniques that are most satisfactory for acclimatization address the changes towards lower relative humidity, higher light leaves, autotrophic growth, and septic environment that are characteristic of the greenhouse and field.

The process of acclimatization can begin while the plantlets are still *in vitro* by subjecting the plantlets to lower relative humidity, which can be achieved by using a desiccant inside the culture vessel. Bottom cooling has also been adopted to reduce the relative humidity in the head space of the container (Maene and Debergh, 1987).

Another method of acclimatizing plantlets while *in vitro* is to uncap the culture vessel several days prior to removal of the plantlets. Uncapping can be done in stages, first by loosening the lids for a day or two, then by partially removing the cap for another day or two, followed by complete removal. This gradual approach is especially helpful if the vessels are in a room with low humidity or without humidity control. Contrary to what one might expect, contamination of the sucrose-containing medium generally does not become problematic unless plantlets remain in the open vessels for more than a week.

23.7.1. Procedures for Acclimatization

Although specific details of acclimatization may differ among laboratories, certain generalizations can be noted. When plantlets have been removed from the culture vessel, the agar has to be thoroughly washed from the roots because the sucrose and nutrients present can serve as a medium for the growth of potential disease-causing organisms. Pesticides may be phytotoxic to some species of micropropagated plantlets, hence, its use is discouraged during the first two weeks after transplanting.

23.7.2. Humidity

Maintenance of a high relative humidity for the first few days following transplanting is critical for survival of the plantlets. Many commercial laboratories have avoided the use of an automatic mist system for this purpose because mist leaches nutrients, causes the medium to become too wet, allows the plantlets themselves to dry, and creates an environment favourable for the growth of algae, some fungi and bacteria. Fogging frequently has been used for acclimatization because it avoids many of the problems encountered with mist system in spite of it being expensive to install.

23.7.3. Antitranspirants

Several workers have adopted the use of antitranspirants (e.g., ABA) to reduce water loss during acclimatization. Though it substantially decreases the stomatal transpiration, in several cases, it caused a large increase in stomatal resistance on the leaves. The use of antitranspirants, therefore, may be limited.

23.7.4. Light

While *in vitro*, plantlets have been exposed to relatively low leaves of light and their leaves are thin and thus resemble shade leaves. Leaves of micropropagated plantlets placed under too high a light level will become chlorotic and scrotched. Some laboratories increase the light level during *in vitro* rooting. Plantlets generally have been acclimatized for up to four weeks under 90% shade. Following a period of shading, plantlets should be gradually moved to the light level under which they will be grown.

23.7.5. Soil and Containers

The growing medium and the container to which *in vitro* rooted plantlets are transplanted are important for good survival. Inhibitors or dramatic shifts in pH of a medium can adversely affect root growth and thus transplanting success. Plantlets are usually transplanted into a uniform medium that adequately supports the plant, has a suitable pH, is well buffered, reproducible, and sufficiently porous to allow adequate drainage and aeration. Although excellent success has been obtained with peat plugs or small foam blocks, it has been argued that plantlets develop better and grow more rapidly in larger containers.

23.8. CONTROL OF DISEASES

Sanitation and disease prevention is essential for transplanting success during and following acclimatization because the plantlets are susceptible to disease-causing organisms. The common problems are leaf and stem decay caused by various fungi and bacteria. An integrated approach of sanitation and application of pesticides is generally used. Other methods used to ensure sanitation include using disinfested medium, new or disinfested containers and benches, new polythene covers or those that have been properly treated to control pathogens, and clean and disinfested containers.

23.9. CONTROL OF TEMPERATURE

The temperature of the air and growing medium are generally controlled during acclimatization. Adjusting the amount of shading and humidity can aid in temperature control. Cooling can be accomplished by ventilation, fan and pad systems, fog, or air-conditioners. The temperature of the root zone is important to encourage root growth. The medium should be warmer than the air for good root activity.

23.10. LIMITATIONS IN ACCLIMATIZATION PRACTICES BY STUDENTS

A couple of decades back, tissue culture inventions, though hailed as exciting, the poor results of hardening were dampening the research on micropropagation. Hardening of plantlets of different species were difficult to master. The reasons could be many, but the major factor was poor greenhouse environment. Of course, now the scenario is different in that many species could be fairly easily hardened. Nevertheless, hardening is still a dampening technique for students. The only solution for increasing the success rate in hardening is to use a good shoot (a elongated shoot with good number of leaves) for *ex vitro* rooting and hardening and well rooted microshoots for hardening to make sure the micro environment such as humidity, temperature and light is as comfortable as possible to survive for a period of time.

At a college laboratory level, in the absence of hardening units, students can evolve their own simple methods to reduce desiccation by providing a high humidity environment around microshoots under shaded conditions.

Propagator boxes could be a good alternative for small scale or experimental hardening work.

24
In Vitro and *Ex Vitro* Rooting of Microshoots

Once a crop of shoots have been produced, the next stage of root initiation may be carried out *in vitro* or the shoots may be harvested and treated as micro-cuttings under non-sterile conditions. Either way, cultures are usually separated into individual shoots at this stage.

24.1. *IN VITRO* ROOTING

With some species, adventitious roots form readily even during the multiplication stage whilst others produce roots when simply transferred to cytokinin-free medium. However, many species require particular conditions usually involving auxin in the medium. There is considerable specificity in responses to types, levels and combinations of auxins and other culture conditions.

Cytokinins generally inhibit root formation (exceptions are there) and so are not included in the rooting media. There may be a carry-over effect of cytokinin treatments during the previous multiplication stage, so an intermediate step on hormone-free medium may be beneficial.

Apart from weaning of from cytokinin treatments, pre-conditioning of the shoots before excision and transfer to rooting medium may promote rooting. Pre-conditioning may include etiolation or chilling treatments.

Whilst auxins may promote root initiation, they can inhibit subsequent root growth. Therefore, a transient exposure to the hormone may be more effective. There may be no visible sign of rooting on the initiation medium but they emerge on transfer to an auxin free medium.

Rooting of woody species is usually more difficult than herbaceous species. For woody plants, cultures derived from seedlings in juvenile phase may root more readily. A similar effect may also be seen when the cultures are passed through numerous multiplication cycles or subdivisions.

24.1.1. Use of Growth Regulators for *in vitro* Rooting

In some species, it is simply needed to transfer shoot or shoot clusters to a hormone-free medium to induce rooting. Most frequently, however, root formation is inhibited by the cytokinins used to induce shoot multiplication, so that shoots do not form roots *in vitro* until they have been cultured on a medium containing auxin alone. Auxins most frequently incorporated into the media to induce rooting are IAA (0.1 - 10 mg/l), NAA (0.05 - 1mg/l) and IBA (0.5 - 3 mg/l). The compounds can often be treated as alternatives, but sometimes, root formation is much better with one auxin than another. Roots initiated by an auxin may sometimes fail to grow in its presence, and shoot growth may be similarly suppressed. The problem is solved by moving shoots from an auxin-containing medium to one without growth regulators.

24.1.2. Media Effects

Satisfactory rooting can take place on full strength culture media but it is a very common practice to transfer shoots to be rooted from high strength media (such as MS media) to less concentrated one or a dilution of the same medium used for culture multiplication. Dilute media formulations commonly used for root initiation are ¼ - ½ MS media.

Among the microelements, boron is important for root formation.

24.1.3. Use of Activated Charcoal

Shoots of a number of different species have been discovered to root more readily *in vitro* when activated charcoal has been added to the medium, sometimes together with an auxin. Stimulation of root growth in the presence of charcoal has been ascribed to the charcoal adsorbing inhibitory substances, absorbing auxins, or to its effect in darkening the culture medium.

24.1.4. Culture Environment

Root hairs do not normally grow on roots that develop in agar, presumably because of poor aeration. Roots bearing root hairs are produced however, when shoots cultured in liquid media and are formed in still greater numbers when shoots are supported above the medium on filter paper bridges.

Although on intact plants growing in soil, root initiation and growth occurs in darkness, light may influence these processes *in vitro*. Root formation in light is sometimes improved if the cultured material is subjected to a preliminary period of darkness in the presence of an auxin.

Root formation is generally favoured by relatively high temperatures and even plants from cool climates have optima which are much greater than normal soil temperatures.

24.1.5. Protocol for *in vitro* Rooting of Jackfruit (Sathyanarayana and Blake, 1993)

Procedure

- Cultures of jackfruit are multiplied using nodal segments.
- Harvest the microcuttings of 2.5- 3.5 cm length at the base just below a node, approximately 10 weeks after culture.
- Remove the basal leaves leaving 2 top leaves (or three in cases where the leaf sizes are relatively small).
- Place the microcuttings onto the MS media with IBA (10^{-6}M) taking care to see that the base of the shoot with at least one node is firmly inserted.
- Incubate the cultures in growth rooms at 25°C temperature with a 16h photoperiod. (Plate 53)

Plate 53: *In vitro* rooted plantlets of jackfruit (After Sathyanarayana, 1993)

24.2. *EX VITRO* ROOTING

Although it is a common practice to root microshoots *in vitro*, this may result in several problems. Roots formed on microcuttings *in vitro* were often found to be thicker, with thicker roots and underdeveloped vascular systems compared to those that formed in fine sand (McClelland, 1989). Such roots formed *in vitro* frequently died, or collapsed after plantlets were removed from cultures. However, new lateral and adventitious roots formed during acclimatization and the plantlets continued to grow vigorously. Many commercial laboratories do not root microcuttings *in vitro* because it is labour intensive and also expensive. The process of rooting *in vitro* has been estimated to account for approximately 35 to 75% of the total cost of micropropagation. In the pursuit of cutting costs in the micropropagation of plants, the use and improvement of *ex vitro* rooting methods will undoubtedly be important to commercial laboratories who will want to develop their own efficient and cost effective methods.

Rooting shoots after removal from culture produced plants with a higher root : shoot ratio than *in vitro* rooted plantlets. Rooting and acclimatization when done simultaneously can substantially reduce costs of micropropagation and add efficiency to the production schemes. When combining rooting and acclimatization, the same environmental considerations apply as when acclimatizing rooted plantlets. Particular attention must be paid to humidity, light and temperature. Treatment with root-inducing growth regulators may be required prior to transfer to a rooting medium.

The size of the microcuttings can be critical to rooting after removal from culture. In general, microcuttings rooted best if they were 2.5 to 5 cm long. Shorter microshoots show lower percentages of rooting and microshoots longer than 6 cm are thought to have too much leaf area for optimal rooting.

24.3. ANATOMICAL CHANGES DURING ROOTING

Adventitious root production varies tremendously between species, between cultivars and also depending on age and nature of the organ or part. Anatomical events that take place during rooting may not be the same in different species and the origin of root initiation may differ with different species. It has been recorded that sites near the cambium and phloem are favourites for root initiation in most woody species.

A range of possible anatomical events in rooting can be recognized, though sometimes, all these events are not present. Hartmann and Kester (1983) suggested that there are three stages in adventitious root formation:

(*a*) Initiation of root initials
(*b*) Differentiation into root primordia, and
(*c*) Development and macroscopic emergence of the primordia

Lovell and White (1986), based on their review recognized four possible stages:

(*a*) Induction and activation shown by the occurrence of cell divisions.
(*b*) Further division of the cells. These cells are termed primordial initials, though the primordium may not be determined at this stage.
(*c*) Increase in cell number and organization. This determined structure is considered a root primordium.
(*d*) Further increase in complexity and development into root.

Sathyanarayana and Blake (1993) identified the following stages during root initiation in microcutting of jackfruit:

(*a*) Signs of root initiation were visible as early as four days after placing the microcuttings in the medium indicated by cell divisions outside the cambium region.

(*b*) Organized root primordia was clearly seen at 8 days.

(*c*) Roots appeared to originate just outside the cambium. (Plate 54)

In an experiment conducted by the above authors, a detailed study of anatomical changes taking place at the base of the shoots placed in a rooting media starting from zero day till the 12th day has been monitored by studying the hand made sections using fluorescence microscopy. (Refer Chapter 29 on Microscopic and photographic techniques for plant tissue culture for details). The following description with photographs gives a brief account of changes taking place in the process of initiation and growth of adventitious roots in microcuttings of jackfruit (Plates 54 a-h).

Occurrence of some cell division activity outside the cambium region indicating a probable root initiation site was fairly evident as early as 4 days after placing the microcuttings in the rooting medium (Plate a). At 6 days, pronounced cell division activity was observed (Plate b). Fromation of root primordium was clearly evident at 8 days (Plate c). The structure like vascular bundles and lignified tissues were also seen fluorescing greenish yellow. The longitudinal section of the microcutting stained with Acridine Orange suggested that root initiation site may have originated outside the cambium (Plates d and e). The fibres are being pushed by the dome of the organized primordium (e). The root primordium was clearly emerged out from the stem at 10-12 days and grew into the medium. Often callus was seen first prior to the emergence of root primordia pushing through the band of fibres (Plates f-h).

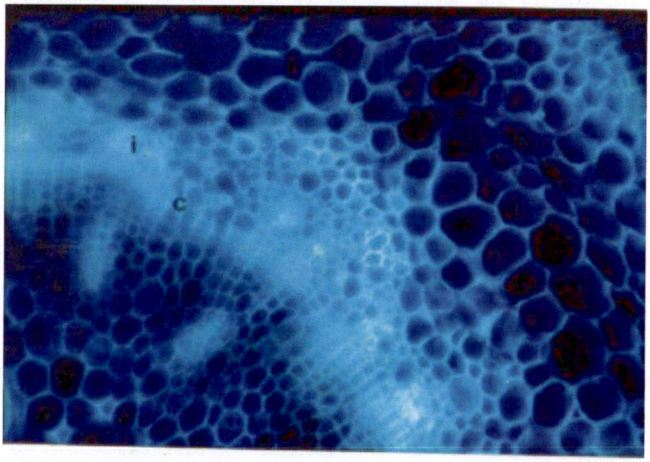

Plate 54 (a-h): Anatomical changes taking place during rooting
54a: TS of the base of a microcutting at 4 days after the rooting treatment showing probable root initial (i) outside cambium (c) (200x) (calcofluor stain)

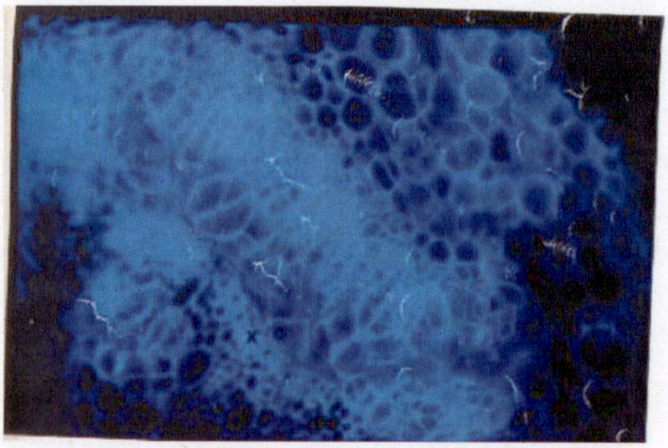

54b. TS of the base of a microcutting at 6 days after the rooting treatment showing pronounced meristematic cell division activity outside the cambium region. (200x), (calcoflour stain)

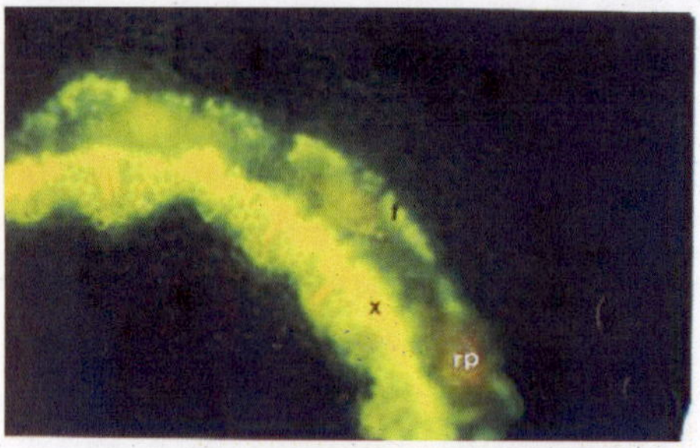

54c: TS of the base of a microcutting at 8 days after the rooting treatment showing root primordia (rp). Note the xylem (x) fluorescing yellow and fibres (f) (100x), (Acridine Orange)

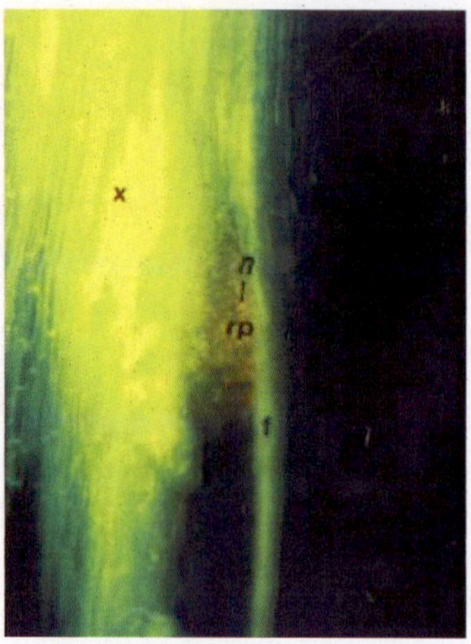

54d: LS through the base of a microcutting at 8 days after the rooting treatment showing the root primordium (rp). Note the fluorescing nuclei (n) stained with Acridine Orange (100x), x = xylem, f = fibres

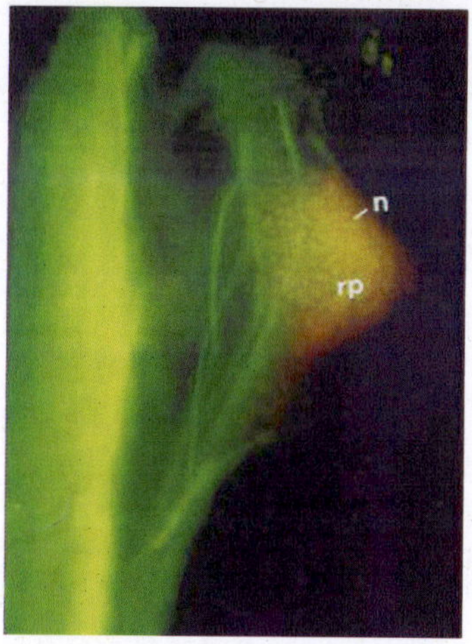

54e: LS through the base of a microcutting at 10 days after the rooting treatment showing emerging root primordium (rp) pushing through the fibres. Note the fluorescing nuclei (n) stained with Acridine Orange (100x)

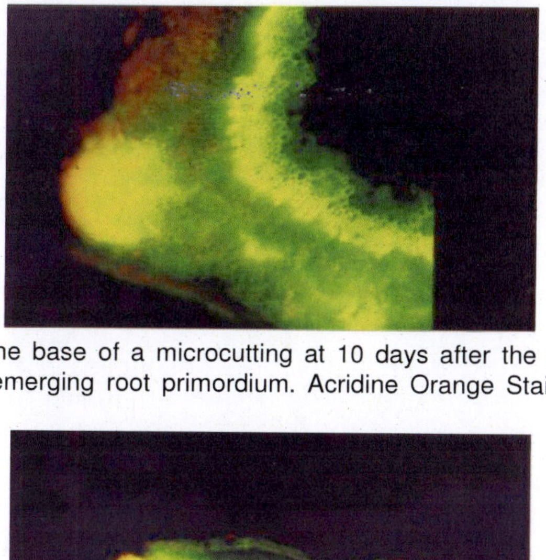

54f: TS through the base of a microcutting at 10 days after the rooting treatment showing emerging root primordium. Acridine Orange Stain (100x)

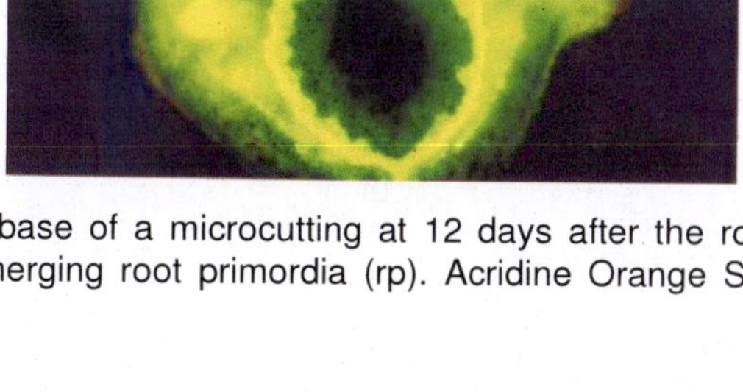

54g: TS of the base of a microcutting at 12 days after the rooting treatment showing emerging root primordia (rp). Acridine Orange Stain (100x)

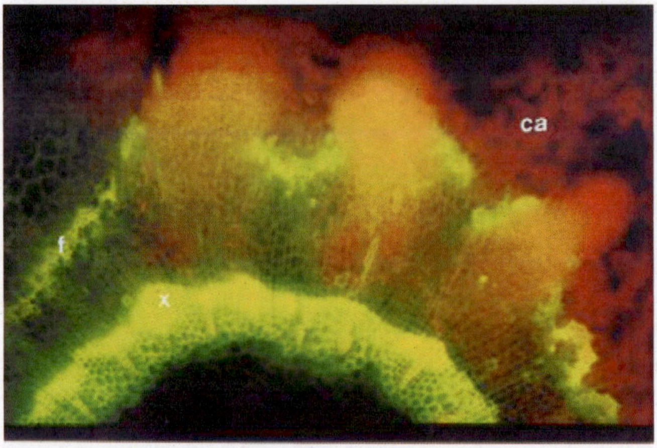

54h: TS of the base of a microcutting (after 14 days) showing the emerged primordia growing out as roots into the medium (intense Orange coloured). Note the callus (ca). (100x), Acridine Orange stain. x = xyless, f = fibres.

24.4. PROTOCOL FOR *EX VITRO* ROOTING OF *IN VITRO* GROWN MICROCUTTINGS OF JACKFRUIT (Sathyanarayana and Blake, 1993)

Procedure

- Select *in vitro* grown microcuttings of jackfruit of 2.5 - 3.5 cm length (from approximately three months after subculture).
- Remove the shoots from the culture tubes and wash off the agar adhering to the base gently by rinsing in water.
- Remove the basal leaves and retain only the top two or three leaves.
- Dip the microcuttings in 0.1% solution of Captan and Benlate for 15-30 seconds.
- Dip approximately 5 mm basal portion of the microshoots in 500 mg/l of IBA for 30 seconds.
- Plant the microcuttings in vermiculite and incubate in the greenhouse with intermittent mist providing a relative humidity of 60-70 per cent.
- Maintain the temperature at 28 ± 2°C and a photoperiod of 12h/16h. (Plate 55)

Note to students/researchers/commercial units : It is always a good idea to give more emphasis on *ex vitro* rooting (rooting microshoots outside laboratory conditions) of any plant species. Being successful at that means you have obtained plantlets which can survive and perform better and also you have saved in the money invested.

Plate 55: *Ex vitro* rooting of jackfruit microcuttings

25
Somaclonal and Gametoclonal Variations

The public concern with the real and perceived safety of Genetically Modified Organisms (GMOs) has forced plant genetcists to reconsider the importance of some older methods of plant improvement including traditional hybridization and the use of somaclonal variation. Somaclonal variation refers to the heritable changes that result from *in vitro* procedures (Skirvin *et al.*, 1994). A small percentage of plants regenerated adventitiously from tissue culture exhibit traits not previously observed. When the variation is heritable it is said to be somaclonal variation. Somaclonal variation has been implicated in changes in characters such as growth habit floral expression, fruit quality, disease and pesticide resistance, the presence or absence of thorns (prickles), and flower colour. Named cultivars have been developed using somaclonal variation protocols (Skirvin *et al.* 1994). It often occurs at higher frequencies than chemical or radiation induced mutation, making it a viable alternative to mutagenesis. Evans *et al.* (1984) coined the term gametoclonal variation for variant plants specifically raised from gametic or gametophytic cells.

Although variation is a serious problem for the nursery people who demand clonal stability, the ability to induce or select somaclonal variation can yield a pool of individuals upon which the selection pressure can be applied for the isolation of unique forms of standard cultivars. Scientists interested in inducing somaclonal variation propagate their plants rapidly *in vitro* on medium supplemented with growth regulators that are expected to stimulate callus and adventitious buds. The amount of variability that can be expected will vary with the clone. If this rate of variability is too low or not reliable, it may be increased by repeated transfers, the use of mutagenic agents, or by applying selection pressure to single cells for stress conditions, ability to resist toxins, or to utilize specific metabolites. The exploitation of somaclonal variation seems especially applicable to older vegetatively propagated cultivars or clones which could be expected to have accumulated large numbers of mutant somatic cells after decades or even centuries of vegetative propagation (Skirvin *et al.*, 1994).

25.1. SOURCES OF SOMACLONAL VARIATION

1. ***Pre-existing variation*:** Cell division is a precisely controlled event that normally yields identical copies of the parental cell. If the control mechanism falters, replicate errors can result in cellular abnormalities.
2. ***Tissue culture induced variation*:** When plant explants are grown *in vitro*, the tissue culture environment itself appears to upset normal controls of cell division and chromosome distribution to result in aberrant cells. For example, McClintok (1984) suggests that the tissue culture environment "resets" or "reprogrammed" plant genomes to yield plants with altered phenotypes.

25.2. FACTORS DETERMINING SOMACLONAL VARIATION

1. ***Type of growth regulator*:** Growth regulators like 2,4-D, 2,4,5-T and BAP enhance the rapid proliferation of disorganized cell. High concentration leads to alteration in the ploidy levels.
2. ***Cultivars*:** The amount of variation arise through *in vitro* is not same for all the cultivars of a species. For example, in case of banana, the rate of variation for dwarf Cavendish is up to 20% and for other cultivars 3%.
3. ***Ploidy level*:** High somaclonal variation has direct correlation with polyploidy or high chromosome level.
4. ***Length of time in vitro*:** One of the important factors. It is must to limit the number of subcultures that can made from an explant. Also if we maintain the cultures, established from a particular explant more than a year, it will lead to somaclonal variation. In case of banana, after 2-3 subculturing, the clone will be replaced with another explant taken from the original mother plant.
5. ***Proliferation rate*:** Normally excessive proliferation will lead to more variation. In case of pineapple by conventional cuttings we can produce 4-5 plants in a year but in case of *in vitro* propagation 30-50 plants per month can be obtained. Here the variation will be more than 50% which is unacceptable commercially.

25.3. GENETIC MECHANISMS / CAUSES FOR SOMACLONAL VARIATION

Although the origin of somaclonal variation is not well understood, there are genetic mechanisms associated with it that have studied in some detail. First of all, we know that as the cell is dividing (*in vitro* or *in vivo*), chromosomes can be distributed unevenly between daughter cells. In addition, the chromosome itself can break during division. Such events can result in somatic mutations that are often called as ***"twin spots"*** on leaves and fruits. Somatic mutations of this sort *in vitro* would be recorded as somaclonal variation.

1. ***Karyotypic changes*:** Can include alterations in the ploidy level (aneuploidy and polyploidy) and chromosomal rearrangements. These alterations have been attributed to late replication of heterochromatin, and nucleotide imbalance as a consequence of culture medium (Lee and Phillips, 1987).
2. ***Point mutations*:** Occur as alterations of one or more adjacent base pairs of DNA (Griffiths *et al.*, 1999). This type of alteration is the least reported among somaclones.
3. ***Gene amplification*:** Is a term used to describe the production of multiple copies of a specific gene in response to a factor in the environment which was absent prior to amplification.
4. ***Transposable elements*:** Are segments of DNA that have the ability to move from one position to another within the genome.
5. ***DNA Methylation*:** Has an important role in DNA regulation and transcription because it affects the binding affinity of enzymes involved in these processes (Wagner *et al.*, 1993). It is known that tissue culture can alter the methylation depending on the composition of the medium (Brar and Jain, 1998). Thus, methylation could be an important cause of somaclonal variation.

25.4. METHODS OF ASSESSMENT OF SOMACLONAL VARIATION

After employing various techniques to create somaclonal variant, is necessary to test the cultures for the genetic stability or possible variation.

1. **Phenotypic parameters**

Quantitative - leaf size, yield, plant height, etc.

Qualitative - branching pattern, flower colours, presence or absence of thorns, disease and pesticide resistance, etc.

2. **Physiological parameters** - protein patterns by electrophoresis for an enzyme, or total content, secondary product formation, e.g., Alkaloid content and steroid level etc.

3. **Genetic parameters** - RFLP, RAPD analysis for alterations in DNA segments

- Chromosome number and structure
- Giemsa / C banding, patterns of chromosome—inversions, deletions.

25.5. APPLICATIONS IN PLANT BREEDING

Somaclonal variation and gametoclonal variation are the important sources of introducing genetic variations that could be of value to plant breeders. Single gene mutations in the nuclear or organelle genome usually provides the best available variety *in vitro*, which has a specific improved character. Somaclonal variations are used to uncover new variants retaining all the favourable characters along with an additional useful trait. For example, Resistance to disease or a herbicide. These variants can then be field tested to ascertain their genetic stability. Gametoclonal variation is induced by meiotic recombination during the sexual cycle of the F1 hybrid results in transgressive segregation to uncover unique gene combinations.

Various cell lines selected *in vitro* and plant regenerated through it prove potentially applicable to agriculture and industry specially resistance to herbicide, pathotoxin, salt or aluminium, useful in the synthesis of secondary metabolites on a commercial scale, etc.

26

Agrobacterium Mediated Transformation

For millennia, cultivators have selectively bred plants in an attempt to induce or enhance a desired trait. Early farmers who settled the banks of the Euphrates learned to better their crops by selecting specific seeds from their harvest for use in the next seasons planting. Similarly, early Americans also selectively bred plants and over a thousand years led to the production of modern corn (maize) from its ancient ancestor, teosinte. Although this was an effective practice, the process usually took hundreds of years and was strictly limited to the alleles present in the same species, or closely related species populations. This need to control and predict the amount of plant food has been a fundamental motivator in developing new plant technologies. Since the advent of recombinant DNA technology in the 1970s, the last twelve years have shown remarkable advances in plant genetics and the rise of plant transformation. These new advances, both biological and mechanical have given researchers the ability to express foreign genes, in desired plant hosts, essentially breaking the plant species barrier.

This powerful technology is of great importance to many aspects of human society. It has opened the door to molecular farming, a novel system for genetic research and new crops adapted to our civilization. The ability to feed our vastly increasing population may very well rest in the hands of plant geneticists. Currently, the practical success of plant genetic manipulation can be seen in agricultural products such as the Newleaf potato, which expresses a natural insecticide and the Flavr Savr tomato in which ripening is inhibited until exposure to ethylene. With many more products on the way, its clear that plant transformation will continue to have a profound impact on plant genetics and global agriculture.

Although plant manipulation, though selective breeding has been around for a long time it was the discovery of plant tissue culture that fuelled modern plant genetics and in turn the development of plant transformation procedures.

The ability to culture plant tissue *in vitro* was first reported at the beginning of this century (Haberlandt, 1902). However, it was not until the 1950s that researchers began to consider the potential applications in industry and research.

Plant transformation has been defined as the stable incorporation and expression of foreign genes into plants. In 1983, Robert Fraley and colleagues reported the successful incorporation of a kanamycin resistance gene into the tobacco genome. Within a year other researchers reported similar accomplishments (De Block *et al*., 1984; Horsch *et al*., 1984; Paszkowski *et al*., 1984), marking the beginning of plant transformation. Since then, successful transformations have been described for over 120 species in 35 different families and the number is increasing every year.

The process of plant transformation is currently a long complex multi-stage procedure involving three general phases. Phase 1 involves the selection and application of a delivery system, which incorporates your DNA of interest into a viable host cell. Phase 2 involves integration, expression and selection of your DNA insert, indicating proper integration and expression of your gene product. Finally, phase 3 involves the recovery of a viable transgenic plant. In many cases, recovery involves tissue culture and regeneration.

26.1. DELIVERY SYSTEMS

A delivery system involves the use of several technologies and requires a vector, which has the capacity to shuttle isolated genetic fragments into a viable host cell. The efficiency of a given delivery system is dependent on the species to be transformed. Researchers have identified which delivery system or combination of delivery systems is optimal for a particular organism.

There are three main classes of delivery systems, biological, physical and alternative, which can be used in transformation protocols. Each system has its advantages and limitations, making it essential for the continuous development and modification of these systems.

26.2. BIOLOGICAL DELIVERY SYSTEM

The search for microorganisms that could be used as vehicles for DNA, lead researchers to the genus of *Agrobacterium*. They noted two species that could be used as a transformation delivery system. *Agrobacterium tumefaciens* and *Agrobacterium rhizogenes* both had the unique ability to transfer portions of their DNA into the genomes of plants. The mechanism by which they transfer DNA is the same, however, the differences in the resulting phenotypes are attributed to the presence of a Ti plasmid in *Agrobacterium tumefaciens* and the

Ri plasmid in *Agrobacterium rhizogenes*. The Ti plasmid DNA induces host plants to grow tumorous masses whereas the Ri plasmid DNA leads to the abundant proliferation of roots. *Agrobacterium tumefaciens* is capable of infecting almost any plant tissue whereas *Agrobacterium rhizogenes* can only infect roots. In its uniqueness, *Agrobacterium tumefaciens* has become the primary choice for many plant transformation experiments.

Agrobacterium tumefaciens (*Agrobacterium*) is a soil dwelling pathogenic bacteria first described in 1943 (Braun 1943). This bacterium has the ability to cause tumorous growths in which the bacteria propagate in the infected plant host. *Agrobacterium* first caught the attention of molecular biologists in the 1970s. It was discovered that its oncogenic activity was due to the presence of a large tumor inducing (Ti) plasmid and tumor formation resulted from its incorporation into the host's nuclear genome. Some strains of *Agrobacterium* are capable of infecting almost all wounded dicotyledonous plants, which makes it a perfect candidate as a transformation vector. The double stranded DNA Ti plasmid encodes the genes responsible for tumor growth. Upon entering the host cell through wounded tissue, a region of the Ti plasmid is excised and incorporated into the host genome. The integrated DNA comprises approximately 25 genes, some of which cause the host cell to overproduce, or to become sensitive to the plant growth hormones auxin and cytokinin. The host cells rapidly proliferate as a neoplastic growth and is commonly referred to as the crown gall tumor.

26.3. THE Ti PLASMID AND TUMORGENESIS

The Ti plasmid of *Agrobacterium* is a large, circular double stranded DNA molecule (T-DNA) of approximately 200 kb, which exist as an autonomous replicating unit. The plasmids are maintained within the bacteria and only a specific region (T-region) approximately 20 kb can be transferred from the bacteria to the host. To accomplish this transfer the Ti plasmid contains a series of genes that code for its own replication, excision from the plasmid, transfer to the host cell, incorporation into the host genome and the induction of tumor formation.

26.3.1. Host Recognition and Binding

Agrobacterium can detect and migrate towards injured plant cells through the detection of chemical signals leaking from the wounded plant. This detection process is referred to as chemotaxis. *Agrobacterium* can recognize plant compounds such as acetosyringone, sinapinic acid, coniferyl alcohol, caffeic acid and methylsyringic acid which induce the bacteria's virulence. To begin the infection process, *Agrobacterium* must bind itself to the host cell. This binding is achieved by a group of genes located within the bacterial chromosome. The precise mechanism for attachment is still unknown. However, it is proposed that these bacteria anchor at the site of injury, by the production of cellulose fibrils. The fibrils attach to the cell surface of the plant host and facilitate the clustering of other bacteria on the cell surface. It is believed that this clustering many help the successful transfer of T-DNA. Once bound to the host, the bacterium is free to begin the processing and transfer of the T-region.

26.3.2. Activation of the *Vir* Genes

The genes that control the excision, processing and transfer of the T-region are called the virulence (*vir*) genes. The 30kb *vir* genes are located outside the T-region on the Ti plasmid and code for proteins, some of which bind directly to the T-DNA. There are at least six *vir* operons (*vir* A, -B, -C, -D, -E, and -G) which mediate the actual transfer. The *vir*A and *vir*G are required for overall *vir* expression indicating they are *vir* master control genes. *Vir*A is autophosphorylated when activated by phenolic compounds released from the injured plant. *Vir*A phosphorylates *Vir*G, which activates the transcription of other *vir* genes. Furthermore, *vir*A and *vir*G mutants lack T-region excision ability providing evidence that they are the master control genes.

26.3.3. T-Region Processing

To transfer the T-region of the Ti plasmid to the host cell, the T-region must be processed such that it is excised from the plasmid and directed to the host. The T-region is flanked by almost identical left (LB) and right (RB) 25bp border sequences, which act as recognition excision sequences. Chemical signals from the injured plant cells (such as acetosyringone) signal bacterial enzymes to excised the T-region from the Ti plasmid. Once removed from the plasmid the T-region, under control of the *vir* genes, is directed into the host cell. Upon activation of the *vir* genes, several changes occur within the T-region. The *Vir*D1 (15kDa) and *Vir*D2 (56kDa) proteins complex and exert single strand endonucleolytic cleavages between the third and fourth base of the bottom strand of the 25bp border repeats. Thus, a single stranded 5′ to 3′ fragment (T-strand) from the T-region is generated moving from the right-bordered repeat to the left-bordered repeat. The remaining single strand portion of the Ti plasmid is copied over by DNA polymerase and ligated together rendering an identical intact Ti plasmid.

26.3.4. Formation of the T-Complex

Once the T-strand is synthesized, it remains as single strand linear DNA and has many hurdles to overcome to reach its final goal, insertion into the nuclear genome. The T-strand must first be protected from degradation by bacterial endo- and exo-nucleases, directed through the bacterial and plant cell walls and incorporated into the host genome.

Protection from endo- and exo-nucleases is thought to be mediated by the *Vir*E2 protein by binding the T DNA, preventing degradation. High abundance of *Vir*E2 during T-region activity suggests that it provides protection by coating the T-strand DNA. It is also thought the *Vir*E2 may also aid in the separation of the T-Region duplex by binding the total length of the T-strand. As mentioned above, the T-strand is produced with 5′ to 3′ polarity. The polarity is maintained through transfer and integration via a 5′ pilot protein.

26.3.5. T-Complex Transfer to the Plant Cell

Once properly packaged, the T- complex transfer is mediated by several proteins and is thought to be similar to bacterial conjugation. The *Vir*B proteins, in complex with other proteins form a 2nm channel with which the T-complex can pass into the plant cell. The hydrophobicity of the channel targets the N-Terminal of the pilot protein out of the cytoplasm, maintaining the polarity of the T-complex. The presence of the *Vir*B proteins is essential for T-DNA transfer. Within the channel complex, *Vir*B4 and *Vir*B11 have been shown to have nucleotide-binding, ATPase and protein kinase sites, which actively translocates the T-complex into the host.

26.3.6. Integration into Host Genome

Once inside the plant cell, the T-complex must be taken up through the nuclear membrane, as the T-complex is too large to diffuse through the nuclear pores. It is theorized that the passage of the T-complex into the nucleus is aided by host cell mechanisms and *Vir* proteins.

The mechanisms of integration into the host genome are still relatively unknown. However, it is known that the T-complex polarity is maintained upon integration. It is believed that the *Vir*D2-bound 5′ end of the T-complex may help unwind the host DNA and form a gap. Next, the 5′end of the T-complex is bound to the gap in host DNA and 3′ end is incorporated down stream. The plant repair and recombination mechanisms may then replicate the DNA using the T-strand as a template and ligate the ends together.

26.3.7. Tumor Formation

The oncogenic activity of T-region resides in the functioning of a number of genes. The T-region carries a gene for an opine-synthesizing enzyme (*O*). Opines are an amino acid derivative that when degraded by enzymes, coded for by genes located outside the T-region on the Ti plasmid, produce nutrients used only by the bacteria and not the host. As well, the T-region contains a group of three genes the code for plant growth hormones (*onc*). The *onc* genes comprise two genes (*iaaM* and *iaaH*) coding for enzymes involved in auxin synthesis and another gene (*iptZ*) that codes for an enzyme that catalyzes the synthesis of cytokinin. Once activated, these two hormones cause the infected cells to divide and allow neighbouring cells to become infected. Within a few weeks, the rapidly dividing infected cells produce a yellow cell mass called a crown gall.

26.4. LIMITATIONS OF BIOLOGICAL DELIVERY SYSTEMS

Currently *Agrobacterium* vectors have been used in the successful transformation of many dicots. However, the fundamental limitation of this system is the reduced ability to infect monocots. Several attempts to extend the transformation range of *Agrobacterium* by using different strains or altering the physiology and co-cultivation techniques of both the bacteria and the host have not been promising. This monocot problem focuses attention on the use of alternative delivery systems, such as particle bombardment and protoplast fusion.

The *Agrobacterium* system requires physically injured plant cells that the bacteria can enter. Whether the procedure involves the whole plant or plant culture; such as calli, protoplasts or leaf disks, the processes requires the regeneration of a viable plant (as with plant cultures), or the generation of viable transformed gametes. Thus, the *Agrobacterium* system will not function with plants that cannot be regenerated from culture or tolerate physical injury.

Finally, the *Agrobacterium* vector system relies completely on the functioning of the *vir* genes, which target only the nuclear genome. Thus the targeting of organelle genomes cannot be accomplished by *Agrobacterium* and alternative delivery systems, such as particle bombardment must be employed.

26.5. CURRENT DEVELOPMENTS AND TECHNIQUES IN BIOLOGICAL DELIVERY SYSTEMS

Since 1983, continuous research and development have produced remarkable advances on the *Agrobacterium* vector system. The majority of advances in *Agrobacterium* vectors have been in the integration region of the binary plasmid. However, modifications have also been made to the tissues that are infected and the method of wounding. Modified viruses such as the tomato golden mosaic virus (TGMV), whose viral coat protein genes were replaced by an *NPTII* resistance gene, can be used to enhance integration. Once inside the genome the virus has the ability to propagate and increase its copy number. The removal of the coat protein genes inhibits the production of new viable virus thus leaving replicated copies of the viral DNA insert within the nucleus. Similar results have been obtained using other viral vector combinations such as the *Agrobacterium*-geminivirus vectors and *Agrobacterium*-cosmid vectors. The Ti system inserts only one T-DNA segment per bacteria into the host's nuclear genome. However, it should be noted that several bacteria could infect the same host cell, effectively increasing the T-DNA copy number in the genome. Thus, both self-replicating viruses and multiple bacterial infections allow for the increased copy number of the DNA of interest.

A recent advance in vector construction for high molecular weight DNA, involved the generation a bacterial artificial chromosome (BAC) for use in conjunction with the *Agrobacterium* binary vector strategy. The BAC can accommodate inserts between 100-180kb and was chosen over the more commonly used yeast artificial chromosome (YAC), as BACs are much easier to construct, screen and maintain than YACs. The BACs have shown stability in both *E. coli* and several strains of *Agrobacterium* making a practical tool for library construction and positional cloning.

The wounding of plant tissue is a critical step in *Agrobacterium* mediated transformation, as it allows the bacteria to enter the host cell. It is thought that this step is a major factor in *Agrobacterium*'s poor ability to infect monocots. Traditionally, leaf disks or cuttings have been used but new technologies make possible to transform almost all plant tissues. A recently developed technique, sonication-assisted *Agrobacterium*-mediated transformation (SAAT) greatly increases the penetration of bacteria in the host cells. The technique involves brief periods in which the target tissue is subjected to ultrasound in the presence of *Agrobacterium*. The treatment produces small uniform fissures and channels in the cells allowing the bacteria easy access into the cytosol. The SAAT has the potential for uniform transformation of meristematic tissue, as previous techniques produces scattered infiltration. Furthermore, the SAAT procedure produced a 100-to1400-fold increase in gene expression in a variety of tissues and organisms.

Another wounding technique, vacuum infiltration, has also been shown to improve *Agrobacterium* mediated transformation. Concentrated solutions of *Agrobacterium* were applied *in planta* and infiltrated the tissue under a vacuum. It is thought that the negative pressure forces the bacteria into the host cells, thus increasing the susceptibility of the host. Furthermore, the vacuum process stimulates the plants wound response, again increasing the virulence of the bacteria.

Another improvement to the *Agrobacterium* system involves the stimulation of the *vir* genes. As mentioned above, the *vir* genes are responsible for the processing, transfer and integration of the T-DNA. These genes and thus the virulence of *Agrobacterium* are stimulated by compounds secreted from the wounded plant tissue. It has been shown that these compounds, such as acetosyringone greatly enhance transformation and are now routinely added to transformation experiments.

26.6. INTEGRATION, RECOMBINATION AND EXPRESSION

Phase 2 of the plant transformation process begins with the successful incorporation of the foreign DNA into the host cell. Depending on the type of delivery system employed, the DNA may already be integrated into a genome. If not, the foreign DNA must undergo recombination or direct insertion into the host genome. Furthermore, there are three different types of genomes in the plant cell, which require different types of insertion. The success of the vector integration is determined by the survival of the host through the demonstration of the selectable markers and the expression of the gene product from the foreign DNA insert.

26.7. SELECTABLE MARKERS AND REPORTER GENES

Transformation vectors are constructed with a reporter gene or a selectable marker, which indicates successful incorporation and expression of the vector genes. The reporter gene or selectable marker gene is fused to the same plant promoter that is fused to the foreign gene of interest. Commonly, antibiotics and herbicides are used as selectable markers. During the regeneration process of the plant transformation procedure, transformed host cell or tissues are propagated in the presence of a substance, which normally inhibits the growth of the plant cells. The expression of the selectable marker gene provides resistance to the inhibitory growth substance, allowing for preferential regeneration of only transformed cells. Reporter genes do not induce survival on a toxic medium but code for a gene product that has an easily detectable phenotype.

26.8. ANTIBIOTIC MARKERS

There are several commonly used antibiotic markers. One of the most common markers is the neomycin phosphotransferase type II gene (*NPT II*) which confers resistance to aminoglycosides (such as kanamycin, paromomycin and geneticin) and was isolated from the bacterial transposon *Tn5*. Another, is a gene isolated from *E. coli*, and induces resistance to the hygromycin toxin. The gene for gentamicin-3-N-acetyltransferase confers resistance to gentamicin and finally the gene for aminoglycoside-3′-adenyltransferase confers resistance to streptomycin and spectinomucin. The most widely used antibiotic marker is kanamycin and was used in the first transformation experiments.

26.9. HERBICIDE MARKERS

Herbicide markers function in the same way as antibiotic markers. There are three commonly used herbicides in plant transformation. The first is phosphinothricin acetyltransferase, which confers resistance to the herbicide phosphinothricin. The second uses a mutant form of the 5-enolpyruvyl-shikamate-3-phosphate synthase (EPSP synthase), which is not susceptible to glyphosate. Non-transformants survive but have an altered phenotype. Finally, the use of acetolactate synthase (ALS) confers resistance to sulfonylureas.

26.10. REPORTER GENES

Unlike selectable markers, reporter genes do not confer resistance to a drug but produce a gene product that has a distinct and easily detectable phenotype. In some transformation experiments, the gene of interest can also act as a reporter if the gene product has a detectable phenotype. The phenotype of the reporter gene should be quick and easy to detect. With most reporters successful transgene expression can be determined through a visual assay. Once a positive visual assay is observed, vector integration can be confirmed by Southern blot hybridization analysis of the host genome. The reporter gene assay is an invaluable step as it eliminates many of the transformants to a select number of possible successful integrations. The two most common reporter genes used in plant transformation are, the β-glucuronidase gene (GUS) and the luciferase gene (LUC). These two non-plant genes are fused to viral or plant promoters and incorporated into the delivery vector.

The GUS reporter gene was isolated from *E. coli* and was shown an effective reporter for plant transformation. The treated plant tissue is transferred to a medium containing 5-bromo-4chloro-3-indoyl-1-glucuronide, which is a substrate for the β-glucuronidase enzyme. Cleavage of the substrate by β-glucuronidase produces blue coloured product. Therefore, any transformed tissue appears blue. One drawback is that some tissues express a basic level of GUS, which can lead to background. An advantage to the GUS reporter is that it can be quantitated by flurometric and spectrophotometric analysis.

The LUC gene was isolated from fireflies. The LUC enzyme requires ATP and luciferin, as a substrate, to produce light. The light emissions are a by-product of the breakdown of luciferin and do not harm the plant tissue. Again, the LUC gene is fused to a viral or plant promoter and incorporated into the delivery vector. Tissues that have incorporated the vector and thus the LUC gene will emit a yellow green glow. Depending on the levels of expression, LUC expression can be visualized under dark-field microscopy or by the naked eye and can be quantified with a luminometer.

Another reporter gene commonly used in the transformation of cereals is anthocyanin. Upon activation, this reporter system produces a reddish-purple pigment in transformed tissue.

26.11. RECOVERY OF VIABLE TRANSGENIC PLANTS

This phase of the plant transformation protocol involves the selection of successful transformants, regeneration and propagation of the new transgenic plant. This is the most time consuming and in some cases, the rate-limiting step of this procedure as it requires a variety of tissue culture techniques. The ability to culture and regenerate plants differs between species. Thus, tissue culture limitations have prevented some species from being transformed. Transformation protocols will use either cell culture, callus culture or *in plantae* propagation of transformed cells. After the delivery system has been administered, successful transformants are selected using selectable markers or reporter genes. Individuals exhibiting successful integration and expression are transferred into regeneration culture conditions.

26.11.1. Selection

There are three stages involved in selecting the plants that have undergone proper vector integration. The first stage is the visual assay. As discussed above, there are several methods to visually determine the possible candidates that have the vector DNA incorporated correctly. This visual assay screens out the majority of the individuals, leaving only a few to undergo the second stage of selection. The possible candidates are subjected to a polymerization chain reaction (PCR) screen or Southern Blot hybridization, both of which use probes specific for the integrated foreign gene. The third stage of selection involves the confirmation that the new transgenic plant possess the desired phenotype. This is accomplished by testing a variety of tissues from the plant to see if the protein product from the foreign gene is present. *In vivo* and *in vitro* studies can then be conducted to confirm whether the transgenic gene product is performing as desired.

In transformation experiments that use *in plantae* treatments of meristems require an additional step. It is difficult to select for cells in the meristem, and furthermore the meristem will contain a chimera of transformed cells, which is not useful as a transgenic plant. Cells in the meristem will eventually give rise to gametes that can then be germinated in the presence of a selection medium. This added step results in the selection of a non-chimeric transgenic plant.

26.11.2. Regeneration

The whole transformation process is dependent on the ability of the species and tissue to produce totipotent cells that can be induced to regenerate into a complete viable plant. This regenerative ability is a necessity for all the different transformation systems.

In vitro callus culture is the most commonly used totiopotent tissue from which whole plants can be regenerated. Callus is defined as an amorphous mass of loosely arranged thin-walled parenchyma cells arising from the proliferating cells of the parent tissue. Depending on the species, callus can be induced from a variety of tissues, including leaves, shoots and roots. The callus is initiated by culturing the excised tissue in a growth medium containing a growth hormone (auxin), such as 2,4-dichlorophenoxyacetic acid (2,4-D). As an example, leaf discs would be cut from a plant and placed in a nutrient rich medium. An appropriate delivery system, such as *Agrobacterium* or microprojectiles would be applied to the tissue. After the treatment, the tissue would be transferred to new nutrient rich medium containing 2, 4-D and the appropriate selectable or reporter gene substrate. Depending on the protocol, the treatment can be administered to the tissue or the callus. Once selected for integration, the callus is transferred to new medium containing regeneration hormones, a balanced ratio of auxins and cytokinins. Once both shoots and roots have formed, the plants are transferred from the culture medium and potted in soil, where it is grown to maturity.

Using the *Agrobacterium* leaf disc transformation protocol, the whole transformation procedure can take 12-20 weeks, involving a regeneration time of 10-16 weeks. Using the microprojectile callus transformation system, the whole procedure requires 24-36 weeks, of which 8-12 weeks is required for regeneration. Plant transformation is long and involved procedure. However, its time commitment is completely defined by the tissue culture and regeneration techniques.

26.11.3. Reproduction

Plant transformation cannot be considered a success unless the plant can produce fertile offspring containing a copy and expressing the foreign gene. Producing a single non-fertile transgenic plant is of little value to researchers. Infertility may indicate genetic damage because of the vector integration. Thus, the transgenic plant must be able to propagate itself and the foreign DNA. Gametes can easily be screened to confirm that the foreign DNA was transmitted and is functional, from the parent to the F1 generation.

26.12. THE FUTURE OF PLANT TRANSFORMATION

Plant transformation is a remarkable achievement. It celebrates the culmination of several plant biochemical physiological and genetic technologies. Currently, plant transformation technologies come in many different forms and are a long and expensive process. However, continued research is devoted to improving these techniques. These technologies are sufficiently developed, allowing the commencement of novel genetic research and the development of novel transgenic plants expressing a variety of phenotypes. It is clear that plant transformation is both directly and indirectly an integral part of our society. With this new technology comes a new responsibility to produce and release transgenic plants in safe and responsible manner. We need only to look at the environmental damage caused by the uncontrolled release of water hyacinth in the United States to affirm the nature of our responsibility. Nonetheless, plant transformation has the potential to solve many of the problems of limited food and other resources, currently facing our society. Finally, as the production of transgenic plants is now a reality, scientists must extend their knowledge beyond the laboratory and learn to deal with the economic, social and legal implications of plant transformation.

It is now clear that the possibilities of plant transformation are for the most part bound only by the limits of our imagination. We will likely see the emergence of new transgenic plants conferring traits from frost resistance to biodegradable plastics to glow-in-the dark Christmas trees. We will see the production of super crops that can grow in any condition and generate vast amounts of food. We may also see transgenic plants feeding starving nations, cleaning polluted water and decorating our cities. Although the potential to make wonderful and new transgenic plants is clear, patent laws and environmental restrictions will hinder production. A major limiting factor is that plant transformation is that it is integrally linked to money and thus to those who are willing to invest in it. Furthermore, societal opinions can have a great impact on industry and thus will conduct the direction of the market for transgenic plants.

27

PROBLEMS IN PLANT TISSUE CULTURE

27.1. CULTURE CONTAMINATION

Rapid production of pathogen-free plants is a fundamental goal of the micropropagation process. Since the surfaces of plants are normally covered with a diverse bacterial and fungal microflora, the primary objective of Stage I is the complete elimination of contaminating organisms and the subsequent establishment of the sterile explant *in vitro*. This is usually accomplished through surface sterilization of excised shoot tips (or other explants) with sterilents such as alcohol and/or sodium hypochlorite.

Contamination in tissue culture can originate from two sources, either through carry over of microorganisms on the surface or in the tissues of explants, or through faulty procedures in the laboratory. In the course of plant growth and development, many surface and rhizosphere-inhabiting microorganisms may enter the tissues of the plant through natural openings, wounds etc. opportunistically and to varying extents colonize the plant tissues.

Many of the microorganisms that are likely to be present intercellularly in plant tissues will be capable of growth on the plant tissue culture medium, although some may be inhibited by the high salt or sugar concentration of the pH. Adapted organisms may rapidly over run the slower responding plant tissues. Less adapted microorganisms may proliferate in the tissues of the explant availing of more abundant nutrient supply from leakage of the damaged tissue. These microorganisms may cause a 'halo' effect around the explant that may be seen by examining the culture against background light. Contaminants that are subliminal, may not be expressed during the early stages of culture, but may be expressed downstream in production. Such contaminants may affect the productivity both *in vitro* and of the progeny plants. Included in the non-expressed or subliminal contaminants may be latent pathogens of the crop. Both classes may cause economic losses when the microplants are exposed to different nutrient regimes or environments.

27.1.1. Effects of Contaminating Microorganism on *in vitro* Cultures

Economic losses due to contamination can be readily ascribed to overrunning of cultures in Stage I, which may prevent the introduction into culture of specific genotypes, and to laboratory contamination due to system failures, infestation with mites and thrips or poor sterile techniques. The latter are amenable to improvement through improved work practices while, the former should not result in high loss as the inputs to Stage I are relatively low. More serious losses can be attributed to losses late in Stage II or in subsequent stages when inputs are high in labour, material, etc. Large-scale losses of this kind are most likely to be a consequence of poor monitoring of production combined with the clonal multiplication of subliminally infected stock (Plate 56).

Plate 56: Region near the shoot is free of contamination due to production of allelochemicals (After Sathyanarayana, 1993)

27.1.2. Organisms Associated with Plant Surfaces

The main organisms associated with plant surfaces are fungi, yeast, bacteria and mollicutes (mycoplasmas, spiroplasmas and related organisms). These microorganisms inhabiting plant surfaces derive nutrition from the plant. Thus, areas where food is abundant like nectaries, wounds, senescent tissue etc. are more likely to support high populations. Bacterial and fungal contaminants can also build up on insect residues. Further, crevices, dense hairs, and mucilaginous surfaces can all harbour microorganisms. Many of the potential facultative endophytes may enter the host tissue through natural openings or wounds. The consequence of such entry into host tissue is that such microorganisms may escape the effects of surface sterilents.

A second class of endophytic microorganisms is the intracellular forms, which include the viruses and viroids. Intracellular pathogens are usually vector transmitted although some may spread by contact between infected and healthy plants.

27.1.3. Screening Plants for Contaminating Microorganisms

Once the explants have been established *in vitro*, it is essential that cultures be indexed (screened) for the presence of microbial contaminants before being multiplied in Stage II. The common methods used in the detection and identification of microorganisms in tissue culture is given in Table 11. Besides increased culture mortality, the presence of latent infections can result in variable growth, tissue necrosis, reduced shoot proliferation and reduced rooting. Indexing is accomplished by inoculating tissue sections onto enriched selective microbiological media that will promote the growth of bacterial, fungal or other contaminants. Since secondary contamination can occur as a result of poor aseptic technique or contaminant vectors such as mites, it is important that Stage II cultures also be routinely indexed.

In reality, many laboratories use only visual methods to screen culture contamination (the so-called eyeball determination). There are several important reasons why visual screening for microbial contamination is not adequate. First, visible growth of specific contaminants may be reduced in plant culture media. Certain contaminants such as *Methylobacteria* have evolved close biochemical relationships with the plants and do not grow independently. Finally, bacteria, fungi and other contaminants often persist as endophytes within plants cultures that appear visually contaminant-free. Numerous procedures have been developed to screen for cultivable contaminants in culture. Knauss (1976) has developed procedures which screen for the presence of systemic bacteria and fungi by culturing 0.5-1.0 mm thick internodal and nodal stem cross-sections from Stage I plantlets in enriched broth media that promote microbial growth (Table 11). This procedure has been effective in establishing cultures of foliage plants such as *Dieffenbachia* that are free of pathogens such as *Xanthomonas dieffenbachiae* and *Erwinia chrysanthemi*.

Procedures used to routinely screen for cultivable bacteria and fungi in our Stage I and Stage II cultures are outlined below:

Table 11: Methods used in the detection and identification of microorganisms in tissue cultures

Test	*Application*
Non-specific tests	
Culture indexing	Detection of cultivable bacteria
DNA/RNA fluorochrome staining. Ex: DAPI	*In situ* staining of mycoplasmas and related prokaryotes
Leaf dip electron microscopy	Rod morphology viruses, high titre viruses
Gel electrophoresis	Viroids
Specific tests	
Enzyme linked immunosorbent assay (ELISA)	Viruses and bacteria
Protein A latex-linked antiserum test (PALLAS)	Viruses and bacteria
Immunosorbent electron microscopy	Viruses
DNA Probes	All organisms
Rapid diagnostic kits	Bacteria
Fatty acid profiling	Bacteria

27.1.4. Procedures

27.1.4.1. Indexing Media and Preparation

Difficulties often exist in detecting contamination without knowledge of appropriate culture media for the contaminants present. Therefore, three enriched media are routinely used (Table 12). Sabouraud dextrose medium is formulated for the culture and growth of fungi including yeast, mold and aciduric microorganisms. Yeast extract dextrose broth medium is used to stimulate bacterial growth. AC broth is a sterility test medium for a wide variety of microorganisms. These media are manufactured by Difco Laboratories (Detroit, MI) and distributed through Fisher Scientific Supply (Norcross, GA). Both solid and liquid broth indexing media can be used. The solid indexing media are prepared by the addition of 10 g/L agar. Media are prepared without pH adjustment, and dispensed as 10 ml volumes into 20 ml glass scintillation vials covered with autoclavable screw caps. After autoclaving, the agar-supplemented medium is cooled and solidified as 45° slants.

Table 12: Fungal and bacterial indexing media[1]

Sabouraud Dextrose	*Yeast Extract Dextrose*	*AC Broth*
Broth[2]	Broth	
10 Grams Neo-peptone	10 Grams Yeast Extract	20 Grams Protease Peptone No. 3
20 Grams Bacto Dextrose	10 Grams Dextrose	3 Grams Bacto Beef Extract
		3 Grams Bacto Malt Extract
		3 Grams Bacto Yeast Extract
		5 Grams Bacto Dextrose
		0.2 Grams Ascorbic Acid

Notes : 1. Media available through DIFCO Laboratories, Detroit, Michigan.
2. Media also solidified by addition of 10 g/1 agar.

27.1.5. Culture Indexing Procedure

(*a*) *Stage I Cultures*

Cultures exhibiting visible contamination are discarded. Shoots are aseptically removed from the culture vessel and placed on a sterile Petri dish. Roots are excised and thin cross-sectional discs of stem tissue are cut. Each of the three liquid media is inoculated with a single stem disc from the same culture. The tissue is partially crushed before being inoculated into the liquid media. Agar-solidified slants of each medium are inoculated by first gently stabbing the medium at the base of the slant and then dragging the tissue up along the surface, leaving it at the top of the slant. The remaining shoot culture is transferred onto fresh medium. For reference, the scintillation vials are labelled with the same culture number as the transferred shoot culture. Uninoculated vials serve as controls. Inoculated indexing media and controls are maintained in the dark at 22-30°C for three weeks. Shoot cultures that screen negative for cultivable contaminants are indexed again after 4 weeks.

(*b*) *Stage II Cultures*

Cultures that initially screen negative for contamination can become contaminated as the result of poor aseptic technique or the result of mite contamination. All cultures within a transfer block are numbered. Fifteen percent (15%) of the cultures are randomly selected and indexed using the aforementioned procedures.

Expected Results

Rapid clouding of the inoculated liquid media is a positive indication of the presence of cultivable contaminants in the tissue sample. The presence of contaminants on the agar-solidified media is confirmed by development of colonies on the surface of the medium and/or development of a halo in the medium where the tissue sample was stabbed. The uninoculated controls should not display microbial growth. When contamination is detected in one medium, it is detected in the other two media more than 95% of the time. Detection of slower growing contaminants is usually facilitated on agar-solidified media. Although these media will promote the growth of diverse microorganisms, it is important to note that growth of all bacteria and fungi will not be supported on these media. Other media such as Leifert and Waites (1992) detect a wide range of contaminants with the exception of some *Xanthosomonas*, *Methylobacterium* and *Hyphomycrobium* species and all mycoplasma-like organisms. Likewise, George and Falkinham Mycrobacterium Detection Medium TT (George and Falkinham, 1986) has been developed to detect bacteria of the *Mycobacterium avium-intracellulare* and *M. scrofulaceum* (MAIS) group.

27.2. VITRIFICATION

Herbaceous and woody plants propagated *in vitro* are frequently affected by the redundant presence of various cultural factors that lead to metabolic and morphological derangements. Anatomical, morphological and physiological anomalies in tissue-cultured plants have been described by diverse terminologies—vitrification, translucency, hyperhydration, succulency and glassiness. This disorder, which is mainly manifested in the leaves, affect the two major processes carried out by the leaves, namely photosynthesis and gas exchange. Anomalous anatomy is manifested to a lesser extent also in the stems and roots. These disorders in turn impede micropropagated plant establishment *ex vitro*.

The special requirements for shoot proliferation *in vitro*—high humidity, superfluous nutritional factors, both minerals and carbohydrates, high levels of growth regulators and low light intensity, are the major causes found to induce shoot malformation.

27.3. ANATOMY AND MORPHOLOGY OF MALFORMED PLANTS *IN VITRO*

27.3.1. Morphological Characters

Under some cultural conditions, highly abnormal tissues and organs may be detected after a period of culture: shoots or plantlets may become difficult to propagate *in vitro*, their capacity for survival may be severely reduced and many may perish either in the culture vessel, or upon transfer to the external environment. Losses of up to 60% of cultured shoots or plantlets have been known to occur during commercial micropropagation.

The 'glassy shoot syndrome' can occur with various degrees of severity and is not always accompanied by changes in the phenotypic characters described above. Initially, part of the shoot, or just one or two leaves (especially those touching the medium) may be affected. The shoots of cultures demonstrating mild symptoms often grow more rapidly than normal and may show abnormally high rates of axillary shoot formation.

In the most extreme circumstances, hyperhydric shoots have short internodes and their shoot apices appear fascinated. Affected shoots are often swollen and pale green and their leaves have a translucent, watery or glass-like appearance. The leaves may be elongated or strap-like, and are frequently also turgid and brittle, and have a greyish-blue colour. When shoots in this condition do not elongate in a normal fashion, there is a reduction of apical dominance causing bushy clumps of shoots to be formed. This has been noted in *Dianthus caryophyllus, Gypsophila paniculata*, etc. Highly abnormal shoots do not produce adventitious roots, or root poorly.

27.3.2. Anatomical Characters

Hyperhydric tissues generally exhibit reduced lignification and this causes vessels and tracheids to be missing or possess an abnormal form in stem, buds and leaves. As cells also have relatively decreased levels of cellulose, the plastic properties of their walls are changed causing them to be defective and deformed.

The leaves of hyperhydric shoots have a thin cuticle, a thin-walled epidermis and abnormal stomata. There is abundant parenchymatous tissue in hyperhydric leaves and the spongy mesophyll is composed of large cells. Differentiated palisade cells are reduced in number and in many cases, are completely absent; chloroplasts have an abnormal structure and low chlorophyll content. The lack of normal internal organization results in reduction of both growth and the capacity of photosynthesis.

27.4. FACTORS INFLUENCING HYPERHYDRICITY

1. ***The Culture Environment*:** Hyperhydricity occurs in both herbaceous and woody species and young shoots are most likely to become affected. Symptoms are less likely to occur when cultures are growing rapidly. Hyperhydricity tends to be promoted by high temperature, low light irradiance, or by placing cultures in the dark.

Shoot growth in conditions of high relative humidity is probably the most important environmental factor causing the onset of hyperhydricity. Shoots cultured in liquid media, including those supported over a liquid medium on filter paper bridges, are more liable to become vitrescent than those grown on solid media, but the extent of abnormality varies very much with the genotype. The leaves of plantlets or shoots growing on a solid medium sometimes becomes glassy if they grow into the medium or grow along its surface or touch the side of the culture vessel, but of a normal appearance if they grow upright. Shoots and organs also become hyperhydric if they are covered in condensation.

2. ***Organic and Inorganic Components of the Media*:** In addition to the physical state of the medium, various dissolved constituents may affect plant morphogenesis. Water availability in the culture is determined by the water potential and the matrix potential, both of which are regulated by solutes and gelling agents.

The hyperhydric condition often occurs on full strength MS (Murashige and Skoog, 1962) salts, but not on media with less concentrated macronutrients. Under conditions of poor transpiration, levels of ammonium ion in many media are often too high for the growth of normal shoots and have been found to enhance hyperhydricity. An inappropriate concentration of other ions can also induce the hyperhydricity syndrome, for example, a deficiency in calcium ions or potassium ions.

3. ***Growth Regulators*:** In shoot cultures, the induction of hyperhydricity can be influenced by the type and concentration of the growth regulators used. In many species, it is especially likely to be induced by high levels of cytokinins, while lower rates induce the proliferation of normal shoots. Cytokinins are particularly liable to induce the glassy shoot syndrome when they are added to cultures in which there is another sub-optimal factor (such as when an inappropriate gelling agent has been used, or when cultures have been subjected to high relative humidity or some other stress). The cytokinin BAP has been especially noted to induce hyperhydricity when used at high concentration.

Though an unusual observation, auxins can also induce hyperhydricity. The addition of auxins to media containing cytokinins often does increase the proportion of vitrescent shoots.

27.5. PREVENTION OF HYPERHYDRICITY

Hyperhydricity can be prevented by the following ways:

1. Reducing the relative humidity in the culture vessel. The use of a less tightly fitting closure may sometimes solve the problem. A particularly effective technique is to arrange for the base of the culture vessels to be cooled, particularly during the period just before root induction and transfer to the external environment. Reduction in relative humidity shortly after establishment of a culture may be most effective.
2. Increasing the concentration of agar or gelrite in a semi-solid medium, or increasing the concentration of sugar. This will result in a reduced rate of shoot proliferation in shoot cultures.
3. Use of a two-phase culture technique (Adding a layer of liquid medium above semi-solid medium in the culture vessel) instead of the conventional single-phase agar method.

4. Culturing shoots on a porous support wetted with liquid medium, rather than on semi-solid medium.
5. Where shoots or organs are to be grown on a liquid medium, ensure that they are immersed only periodically and not continuously.
6. Decreasing the concentration of ammonium ions in the medium.
7. Adjusting the initial pH of the medium.
8. Decreasing the concentration of micronutrients in the medium.
9. Replacing sucrose with fructose or galactose, or adding reduced glutathione to the medium.
10. Transferring the cultures to a medium without growth regulators. This may permit the formation of new shoots without hyperhydricity.
11. Decreasing the concentration of cytokinins used and/or changing from the use of a particular cytokinin, particularly BAP to another compound with similar activity for one or more subcultures.
12. Culturing tissues with an auxin/cytokinin ratio more appropriate to the plant species.
13. Growing shoots or plantlets under light of higher irradiance.

27.6. TISSUE BLACKENING OR BROWNING

The cut surfaces of many explants start to discolour soon after excision. The explants, or parts of them, frequently continue to darken when they are introduced into the culture vessel. This type of 'blackening' or 'browning' is often associated with wounding. Explants may also produce pigments (pink, red, brown, reddish-brown, black or blue) sometime after excision, or cultures may give rise to them during subsequent growth.

Damage resulting from the production of dark pigments is usually most severe during the initial stages of a culture, and ceases to be a problem once explants have commenced growth. (Plate 57)

Plate 57: Browning of media due to phenol exudation from jackfruit shoots

27.6.1. Factors Affecting Tissue Blackening or Browning

(*a*) *Effect of Genotype*

The extent of the blackening and growth inhibition, which occurs in culture, is genotype-dependent, being especially severe in genera that naturally contain high levels of tannins or other hydroxyphenols. Differences are also found between species of the same genus and cultivars within a species. Different genotypes not only differ in the amount of phenolic substances produced, the substances released also vary in their toxicity, or plants demonstrate different susceptibilities.

(*b*) *Source of Explants*

Young juvenile tissues are often less prone to browning on excision than older ones. Tissues excised from heavily pruned and etiolated shoots of woody plants are often less liable to browning than those taken from adult parts of the plant.

(*c*) *Time of Year*

A further factor influencing the extent of browning can be the time of year at which tissues are explanted. The influence of the time of the year varies according to the plant.

27.6.2. Causes of Browning

Tissue blackening occurs through the action of copper-containing oxidase enzymes (catechol oxidase), often called polyphenol oxidase, phenolase and tyrosinase (monophenol oxidase), which are released, synthesized or presented with suitable substrates and oxidative conditions when tissues are wounded or senescent. Substrates for these enzymes, which vary in different tissues, are commonly tyrosine or *o*-hydroxyphenols such as chlorogenic acid. Enzymes, normally latent on cell membranes and substrates retained within cell vacuoles, come together when cells are injured.

27.6.3. Prevention of Browning

If exudation from explants or cultures is accompanied by satisfactory growth or morphogenesis, there is no need to take precautions. Only when browning of tissues and the medium is associated with poor establishment, it is necessary to try to prevent it. The different approaches suggested are:

27.7. MINIMIZING THE DAMAGE CAUSED TO THE EXPLANT

In some species, the sterilents used to decontaminate explants can be responsible for accentuating the blackening of explants. When blackening presents a severe problem, it can therefore be advisable to experiment with replacing one sterilent solution with another.

27.8. REMOVING THE PHENOLIC COMPOUNDS PRODUCED

27.8.1. Leaching or Dispersal

In species, where black exudates can be a problem, pre-treatment of plant material can help in obtaining viable explants. Firstly, care may need to be taken not to get exudates from surrounding tissue onto the explant during excision. After that, washing to remove the products released from damaged cells is frequently effective. Explants should be rinsed or left in sterile water for 2-3 hours after isolation and before being transferred to culture. Thorough rinsing after sterilization is also necessary to wash away chemicals used for decontamination, which if not properly removed, may provoke the synthesis of phenolics.

27.8.2. Frequent Transfers

Growth of explants can be limited by toxic metabolites even in the absence of obvious blackening. If explants do not show any growth after 3-4 weeks, their chances of survival may be assisted by transfer to fresh medium. Rapid transfer is imperative if the medium around the explant begins to become discoloured or blackened. Browning or blackening is often apparent on a solid medium where exudates are trapped by the agar and become concentrated in the vicinity of the explant. In such cases, the most widely used method of preventing tissue blackening is to subculture explants frequently, placing them on fresh medium after only a few days. The interval between transfers need to be adjusted according to the severity of the problem. In conjunction with moving explants to fresh medium, it is advisable, at each transfer, to cut away tissue, which has started to become discoloured because, if this tissue is left, the damage is likely to spread. This might turn out to be an expensive procedure, which should not be undertaken in a commercial laboratory.

27.8.3. Use of Liquid Media

Explants are frequently less liable to browning if they are cultured initially on a liquid medium in which the phenolics can easily diffuse away.

27.8.4. Adsorption with Activated Charcoal

By adding activated charcoal to tissue culture media, it is sometimes possible to avoid the build-up of phenolic inhibitors. But since activated charcoal can adsorb growth regulators and other components of the medium, it should be used with caution. Activated charcoal can also be toxic to some tissues. Activated charcoal is often added at a concentration of 0.5-3 g/L.

27.8.5. Adsorption with Polyvinylpyrrolidone (PVP)

Several different kinds of PVP have been used to prevent browning of plant tissue cultures, either as rinses for explants or by incorporating into the media. Phenols are adsorbed by PVP by hydrogen bonding, preventing their oxidation and polymerization. PVP may also combine with oxidized phenolics, preventing further oxidations by phenolase enzymes.

- **Modifying the physical environment so that conditions for the production of phenolics are not optimal**

The activity of enzymes concerned with both the biosynthesis and the oxidation of phenols are increased by light. Tissue blackening has been found to be less in explants taken from etiolated stock plants grown in darkness or very low light intensity, or in some case, short days.

Cultures are often initiated more readily if newly explanted tissues are kept in the dark for several weeks. Blackening is usually reduced or prevented by such dark treatment, and sometimes, subsequent growth is improved even though tissue discolouration is not observed. Dark incubation can also be effective when black exudates are produced by a growing culture.

Browning may also be prevented if it is possible to maintain plant material at a slightly lower temperature than normal, while still obtaining growth and/or morphogenesis.

- **Modifying the Redox Potential of the Tissue**

The tendency of dissolved compounds to be oxidized or reduced depends on the oxidation-reduction (redox) potential of the solution. There are several antioxidants which act as reducing agents that lower the redox potential of solutions and are thus effective in preventing the blackening of isolated plant tissues or plant extracts, and it is assumed that they prevent the oxidation of phenols.

Explants from the plants whose tissues are liable to browning, are often washed in a solution of an antioxidant after sterilization, excised on a paper soaked with an antioxidant, and/or submerged in a solution of antioxidant immediately after excision. A solution of ascorbic acid and citric acid (citric acid acts as a chelating agent and delays the breakdown of ascorbic acid) is commonly used for rinsing freshly isolated explants to delay the browning process.

Antioxidants like ascorbic acid can also be incorporated into the tissue culture media, but since it is heat labile, it should be filter sterilized and not autoclaved.

Other antioxidants used include glutathione, L-cysteine HCl, dithiothreitol, etc.

Reducing Phenolase Activity and Substrate Availability

The rate of phenolic oxidation will clearly be less rapid if there is either a reduction in the activity of oxidative enzymes, or less substrate available for oxidation. Ions like copper and iron are particularly associated with the activity of oxidative enzymes. Omission of these ions from the media during the initial stages has been found to greatly reduce browning of explants.

There are certain chelating compounds which can afford a means of interfering with the action of peroxidase enzymes, in which capacity, such compounds act as antioxidants. Dithiocarbamates and 8-hydroxyquinoline are strong Cu^{2+} complexing agents. Other examples of chelating agents include EDTA, Thiourea, etc.

Temporary use of a more acid medium can also reduce tissue browning as Polyphenol oxidase activity is greatest at pH 6.5. Soaking explants in a mixture of ascorbic acid and citric acid also serves the same purpose.

Altering the Composition of the Medium and the Growth Regulators Used

Browning is often less severe on a dilute medium than on a high salt medium such as MS. Several media components have been thought to be responsible for browning. A reduction in the concentration of nitrogen has been effective in preventing browning in many plants. Omission of $FeSO_4$ and copper initially from the media has also been found to reduce blackening of explants. The compounds could be replaced at later stages without any harmful effects.

Carbohydrate is required for the biosynthesis of phenolic compounds, and reducing sucrose concentration can decrease polyphenols within the tissues.

Polyphenol formation from freshly excised explants could be influenced by growth regulators, but compounds which have induced browning in one species are without effect in another. Browning is often associated with the use of 2,4-D, especially if high concentrations are added to the medium.

27.9. DECLINE IN VIGOUR

Shoot cultures of some plants decline and cease to grow and proliferate after a period of time. On some occasions, decline is associated with the production of phenolic substances and in other cases, it may be associated with vitrescence, habituation or the failure of the cultured shoots to become 'rejuvenated'. On other occasions, decline has been traced to some inadequacy or superfluity in the nutritional or regulatory compounds in the medium.

27.9.1. Decline During Initiation

Shoots of some plants show chlorosis, leaf abscission and a lack of growth during their first few months of culture, unless they are transferred frequently and their bases trimmed. Sometimes, subculturing ahead of time can prevent such degeneration.

27.9.2. Decline at Later Stages

The rate at which new axillary or adventitious shoots are formed and grow to a harvestable size, does not always stay constant for long periods. A variation in rate is common in shoot cultures of woody plants, where the rate of multiplication is typically low during early passages, but increases as the cultured material becomes rejuvenated. Sometimes, the rate of multiplication rises to a maximum and then falls back in successive subcultures.

27.10. CYTOKININ HABITUATION

Cultures of some plants have been found to become cytokinin habituated. This results in excessive shoot multiplication in cytokinin-free medium, but the shoots produced are small, and give rise to few adventitious roots and these plants are difficult to be established *ex vitro.* In such cases, subcultures may require the elimination of cytokinin for one or two passages, a reduction in the concentration used, or the replacement of BAP with another compound, such as kinetin.

27.11. SHOOT NECROSIS

Necrosis strictly describes the death of a part of a living organism. Its occurrence on parts of plant tissues in enclosed vessels can readily lead to death of whole culture.

27.12. CAUSES FOR APICAL NECROSIS

(a) Calcium Deficiency

Apical necrosis occurs particularly in shoot cultures, especially those of woody plants. Although there may be other causes, the most common reason is the calcium deficiency in the apices of cultured shoots due to its inadequate absorption from the medium, or non-translocation. In culture vessels, high humidity limits transpiration and hence, movement of xylem-translocated ions and compounds, so that calcium may not reach apical tissues. In greenhouses, tip necrosis may occur if calcium is leached from shoot apices by excessive watering.

(b) Growth Substances

In some cases, the deficiency of natural hormones has resulted in apical necrosis. The leaves also turned brown and tended to drop when the shoots after few subcultures on a cytokinin medium were transferred to a medium devoid of cytokinin to promote rooting.

(c) Subculture Interval

Leaf yellowing and necrosis both increase when shoot cultures are not subcultured at sufficient frequency. If these symptoms are detected, the interval between subcultures should be reduced. This practice will assist in the maintenance of a high rate of shoot proliferation. The interval between subcultures often has to be adjusted between one cultivar and another.

(d) Rapid Water Loss

Shoot apices may sometimes die if plants lose water rapidly during *in vitro* hardening.

(e) Gaseous Pollutants

Gas or alcohol burners, operated with a limited supply of air, emit large quantities of gases such as butane, methane, ethane, propane, ethylene, methanol and ethanol. If such burners are used for sterilizing instruments, or for flaming the mouth of vessels, the concentration of gases within the laminar flow hoods and within vessels can be unacceptably high. Bud necrosis can be caused by the presence of such volatile carbon compounds within culture vessels.

27.13. PREVENTION OF SHOOT NECROSIS

(a) Increasing Calcium Availability

Tip necrosis can often be cured by increasing the concentration of calcium in the medium, or by changing the physical environment in which the cultures are grown so that transpiration is increased.

(b) Reducing Gaseous Pollutants

Gas burners should never be allowed to burn with a yellow flame within laminar flow hoods. Although burners adjusted to give a blue flame still emit high concentrations of carbon dioxide. Gas emissions are eliminated if an electric Bunsen or oven is employed to sterilize instruments.

(c) Use of a Two-phased Medium

It was possible to partially overcome apical necrosis by pouring a layer of liquid medium of the same composition as the underlying layer onto the surface of the semi-solid medium directly after subculture.

28

COMMERCIAL MICROPROPAGATION

Tissue culture methods are used by nearly 600 companies throughout the world to produce more than 500 million plants annually for almost 50,000 varieties of plants. Currently, almost all of these plants are produced by micropropagation. Large-scale propagation through somatic embryogenesis is either uncommon or an exception. Tissue culture is rapidly becoming a commercial method for propagating new cultivars, rare species, and difficult-to-propagate plants. From a few research laboratories several years ago, a whole new industry is emerging. Currently, the demand for micropropagated plants is greater than the supply with some plants. Some growers specialize in only the micropropagation of plantlets, leaving the growing-on to others; many growers are integrating a tissue culture laboratory into their overall operation.

There are two principal pathways for *in vitro* plant regeneration. (*i*) The most common is micropropagation. It involves the suppression of apical dominance resulting in the multiplication of axillary buds. (*ii*) In somatic embryogenesis, cotyledon containing embryos with a root-shoot axis are formed from somatic cells. At the present time there is a much better understanding of the biology of micropropagation than of somatic embryogenesis.

A great deal of fundamental work of biology on the somatic embryogenesis must be carried out before large-scale production of somatic embryos can become as routinely and widely used for micropropagtion. Although somatic embryogenesis has been studied and described in more than 200 plant species during the past 35 years, the technical know-how for the production of unlimited numbers of somatic embryos, which are uniform in size and shape and retain the capacity to give rise to normal plants, is not available for any species.

The micropropagation industry has stagnated during the past two decades largely for two reasons. First, the limited number of individual plants that can be produced and second, the high costs of production. Estimates of labour cost run from 50% to as high as 85% of total production cost. Automation of the propagation process will result in a drastic reduction of labour use, producing virtually unlimited numbers of plants and opening the market to many vegetable, fruit and tree species.

28.1. DESIGNING A LABORATORY

In designing any laboratory, big or small, certain elements are essential for a successful operation. The correct design of a laboratory will not only help maintain asepsis, but it will also achieve a high standard of work. (Fig. 14)

28.2. FACILITIES

Careful planning is an important first step when considering the size and location of a laboratory. It is recommended to make visits to several other faculties to view their arrangement and operation. A small laboratory should be set up first until the proper techniques and markets are developed.

A convenient location for a small lab is a room or part of the basement of a house, a garage, a remodelled office or a room in the head house. The minimum area required for media preparation, transfer and primary growth shelves is about 150 sq ft. Walls may have to be installed to separate different areas.

A good location includes the following:

1. Isolation from foot traffic
2. No contamination from adjacent rooms
3. Thermostatically controlled heat
4. Water and drains for a sink
5. Adequate electrical service
6. Provisions for a fan and intake blower for ventilation
7. Good lighting

Larger labs are frequently built as free-standing buildings. Although more expensive to build, the added isolation from adjacent activities will keep the laboratory cleaner.

Prefabricated buildings make convenient low cost laboratories. They are readily available in many sizes in most parts of the country. Built-in-place frame buildings can also be used. Consideration should be given to the following:

- Check with local authorities about zoning and building permits.
- Locate the building away from sources of contamination such as a gravel driveway or parking lot, soil mixing area, shipping dock, pesticide storage, or dust and chemicals from fields.
- A clear span building allows for a flexible arrangement of walls.

- The floor should be concrete or capable of carrying 50 pounds per square foot.
- Walls and ceiling should be insulated to at least R-15 and be covered inside with a water-resistant material.
- Windows, if desired, may be placed wherever convenient in the media preparation and glassware washing rooms.
- The heating system should be capable of maintaining a room temperature at 70°F in the coldest part of winter.
- A minimum 3/4 inch water source is needed.
- Connection to a septic system or sanitary sewer should be provided.
- Air-conditioning for summer cooling may be necessary.
- Electric service capacity for equipment, lights and future expansion should be calculated. A minimum 100 amp service is recommended.

28.3. GENERAL LABORATORY DESIGN

Cleanliness is the major consideration when designing a plant tissue culture laboratory. Most companies are not aware of their losses from contamination, but estimates run from less than 1% up to 50%. When you consider the high value of the product, no losses from contamination are acceptable. Routine cleaning and aseptic procedures can decrease your losses to less than 1%. Laboratories should have easy to wash walls and floors. Acrylic or urethane epoxy wall paints can be used; cement floors can be painted with an epoxy or urethane floor enamel or have an inlaid linoleum installed. High efficiency particulate air (HEPA) filters or regular furnace filters can be installed over air intakes to the laboratory or on furnaces. If possible, an enclosed entrance should precede the laboratory; sticky mats can be laid there to help collect dirt from the outside, or shoes can be removed.

The traffic pattern and work flow in a laboratory must be considered in order to maximize cleanliness. The cleanest rooms or areas are the culture room, i.e., primary growth room, and the aseptic transfer area. It is best to design these rooms so they are not entered directly from the outside of a building. The media preparation area, glassware washing area, or storage area should be located outside these rooms. The primary growth room and aseptic transfer room should be enclosed with doors leading to each. Traffic through these areas can be minimized by installing pass-through windows. Ideally, the media preparation area would lead to the sterilization area, which would lead to the aseptic transfer room and eventually the primary growth room.

Unusual requirements for electricity and fire safety dictate that power installation be done by professional electricians. Most wiring will require 110 volts. Temperature and fire alarms are to be connected directly to telephone

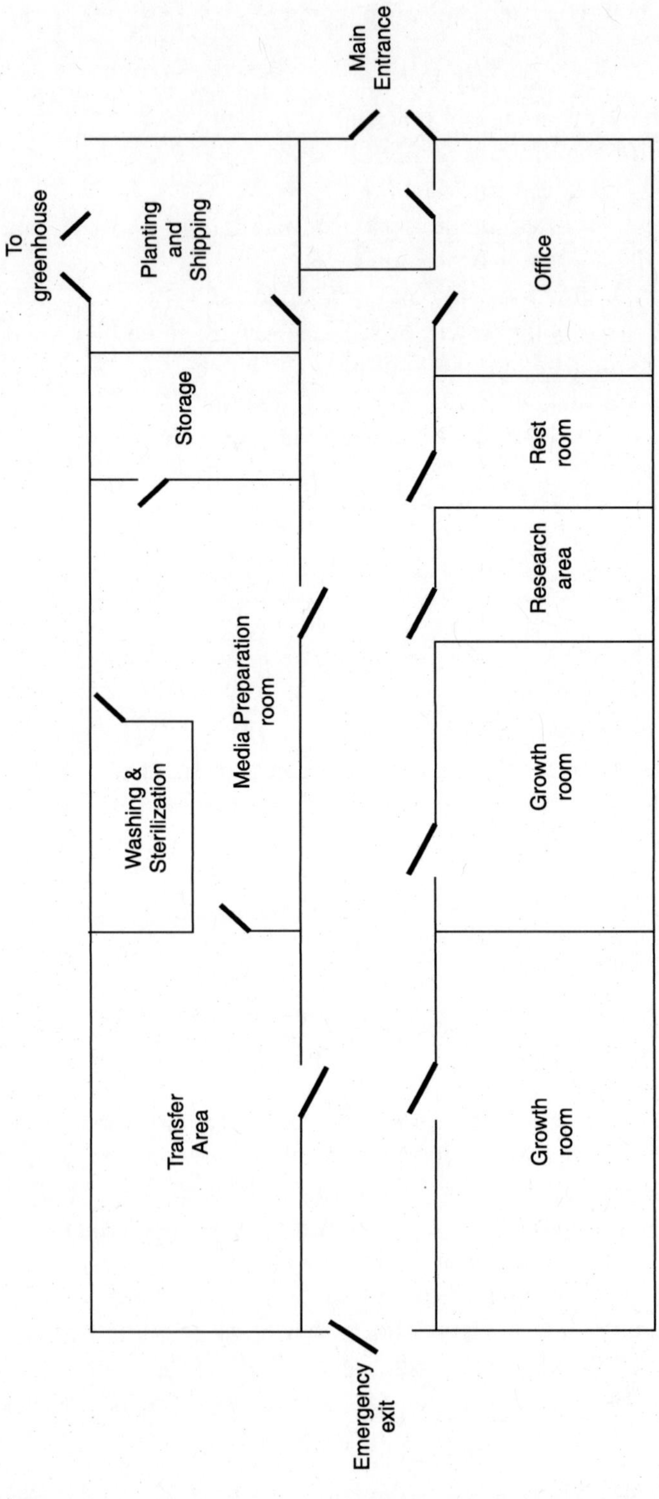

Fig. 14: Floorplan of a commercial tissue culture laboratory

lines to give fast warnings of problems. An emergency generator should be available to operate essential equipment during power outages.

The glassware washing area should be located near the sterilization and media preparation areas. When culture vessels are removed from the growth area, they are often autoclaved to kill contaminants or to soften semi-solid media. The vessels can be easily moved to the washing area if the autoclave or pressure cooker is nearby. Locate the glassware storage area close to the wash area to expedite storage; these areas also need to be accessible to the media preparation area.

The glassware area should be equipped with at least one large sink; two sinks are preferable. Adequate work space is required on both sides of the sink; this space will be used for glassware soaking tubs and drainage trays. Plastic netting can be placed on surfaces near the sink to reduce glassware breakage and enhance water drainage. The pipes leading from the sink can be PVC to resist damage from acids and alkalis. Both hot and cold water should be available with water distillation and/or deionization devices nearby. Mobile drying racks can be stored nearby and lined with cheesecloth to prevent water dripping and loss of small objects. Locate ovens or hot air cabinets (75°C) close to the glassware washing and storage area. Dust-proof cabinets, low enough to allow easy access, can be used in the storage area.

28.4. MEDIA PREPARATION AND STERILIZATION AREA

The water source and glassware storage area should be convenient to the media preparation area. Benches, suitable for comfortable working while standing (34 to 36 inch) and deep enough (24 inch) to hold equipment listed below are essential. Their tops should be made with molded plastic laminate surfaces that can tolerate frequent cleanings.

There are a variety of equipment available for micropropagation laboratories; these equipments are generally located in the media preparation area. The equipment budget will determine the type and amount purchased.

All laboratories need the following basics:

1. ***Refrigerator/freezer*:** This is needed to store chemicals and stock solutions. Small laboratories may find it adequate to use countertop refrigerators.

2. ***High quality water*:** Bottled water can be purchased inexpensively and placed in the media preparation area. Larger businesses may find it economical to obtain distillation or deionization devices; these would normally be located in the glassware washing area. Small, inexpensive, low production Pyrex distillation devices can be purchased by small businesses that want the convenience of a still, but not the cost.

3. ***Balances***: High quality balances are essential for a micropropagation laboratory; this is one area where it is difficult to find an inexpensive substitute. A triple beam balance is useful for large amounts over 10 grams, but a balance that can measure down to 2 mg is essential. Most laboratories have both a microbalance and a less sensitive top loading balance; the latter can be used more quickly and efficiently for less sensitive quantities.

4. ***Hot plate/stirrer***: At least one hot plate with an automatic stirrer is needed to make semi-solid media. This purchase can be eliminated by using a stove and hand stirring the media while it heats; however, the time saved by using a stirring hot plate is worth the money spent.

5. ***pH meter***: This is needed to measure media pH. Some laboratories use pH indicator paper, however this method is considerably less accurate and could severely affect the results.

6. ***Aspirator or vacuum pump***: Aspirators can be easily attached to a water source and used for filter sterilization of chemicals. They are also used to disinfest plant material. Vacuum pumps are faster and more efficient, but also more expensive.

7. ***Autoclave***: An autoclave or pressure cooker is a vital part of a micropropagation laboratory. High pressure heat is needed to sterilize media, water, glassware, and utensils. Certain spores from fungi and bacteria will only be killed at a temperature of 121°C and 15 pounds per square inch (psi). Self generating steam autoclaves are more dependable and faster to operate.

8. ***Optional equipment***: A variety of non-essential equipment is available for tissue culture laboratories; individual's need and equipment cost will determine what can be purchased.

   (*a*) **Microwave ovens** are convenient for defrosting frozen stocks and heating agar media.
   (*b*) **Dissecting microscopes** are useful to have in the laboratory for meristeming, dissecting floral and shoot apices, and observing plant culture growth.
   (*c*) **Labwashers**, or regular dishwashers, can be useful.
   (*d*) **Automatic media dispensers** are helpful when pipetting large volumes of media.

28.5. GROWTH ROOM

Temperature, relative humidity, lighting units, and shelves need to be considered in the culture room. All of these environmental considerations will vary depending on the size of the growth room, its location, and the type of plants grown within it. For example, a small primary growth room located in a cool, North American climate, can be placed in an unheated or minimally heated basement. The ballasts from the fluorescent lights do not need to be separated; rather they can be used as a heating source. Excess heat can be blown out of the growth room and used to heat other parts of the basement or building. In this case, solid wood shelves with air spaces located between shelves are recommended to prevent the cultures on shelves above lights from becoming over-heated. A larger growth room located in an above ground location may need to have remote ballast and/or a heat pump installed. Shelves in a larger growth room could then be glass or expanded metal.

Temperature is the primary concern in culture rooms; it affects decisions on lights, relative humidity, and shelving. Generally, temperatures are kept 76 ± 2°F. Heating can be accomplished by traditional heating systems supplemented with heat from light ballasts or space heaters. Cooling the room is usually a greater problem than heating; low temperature can be obtained by installing heat pumps, air-conditioners, or exhaust fans. Using outside windows to cool culture rooms invites contamination problems in the summer and humidity problems in the winter.

Some plant cultures can be kept in complete darkness; however, most culture rooms are lighted at 1 K lux (approximately 100ft-c) while some are provided with light up to 5 to 10 Klux. The plant species being micropropagated will determine the intensity used. The developmental stage of the plants will also help to determine if wide spectrum or cool white fluorescent lights are used. Rooting has been shown to increase with far-red light; therefore, wide spectrum lights should be used during stage III and cool-white lights can be used during Stages I and II. Automatic timers are needed to maintain desired photoperiods. Reflectors can be placed over bulbs to direct their light. Heat generated by the lights may cause condensation and temperature problems. In addition to previously mentioned procedures, small fans with or without polyethylene tubes attached, can be placed at the ends of shelves to increase airflow and decrease heat accumulation.

Relative humidity (RH) is difficult to control inside growing vessels, but fluctuations in the culture room may have a deleterious effect. Cultures can dry out if the RH of culture room is less than 50%; humidifiers can be used to correct this problem. If the RH becomes too high, a dehumidifier is recommended.

Shelving within primary growth rooms can vary depending upon the situation and the plants grown. Wood is recommended for inexpensive, easy-to-build shelves. The wood for shelves should be exterior particleboard or plywood and should be painted white to reflect the room's light. Expanded metal is more expensive than wood, but provides better air circulation; wire mesh of 1/4 or 1/2 inch hardware cloth can be used but tends to sag under load. Tempered glass is sometimes used for shelves to increase light penetration, but it is more prone to breaking. Air spaces, 2 to 4 inch, between the lights and shelves will decrease bottom heat on upper shelves and condensation in culture vessels. A room that is 8 ft high will accommodate 5 shelves, each 18 inch apart, when the bottom shelf is 4 inch off the floor. The top and bottom shelves may be difficult to work.

28.6. ASEPTIC TRANSFER AREA

In addition to the primary growth room, the aseptic transfer area needs to be as clean as possible. It is preferable to have a separate room for aseptic transfer; this decreases spore circulation and allows personnel to leave shoes outside the room. Special laboratory shoes and coats should be worn in this area. Laminar flow hoods or still-air boxes can be placed in this room and used for all aseptic work. Ultraviolet (UV) lights are sometimes installed in transfer areas to disinfect the room; these lights should only be used when people and plant material are not in the room. Safety switches can be installed to shut off the UV lights when regular room lights are turned on. Surfaces inside the aseptic transfer area should be smooth to minimize the amount of dust that settles. Several electric outlets are to be installed to accommodate balances, flow hoods, bacticinerators, and microscopes.

28.7. THE COST OF MICROPROPAGATION

The cost of micropropagation is influenced by a number of factors, many of which are dependent on the genotype:

1. The capital and overhead cost of the laboratory
2. The cost of labour
3. Material costs
4. The *in vitro* characteristics of the plant to be propagated like how easy it is to establish a culture capable of sustained multiplication, how rapidly and quickly a plant can be multiplied and whether shoots need to be rooted in culture or they can be rooted *ex vitro*.
5. The losses incurred at each stage of the micropropagation process which includes:
    - (*a*) the proportion of cultures lost through contamination
    - (*b*) the proportion of shoots affected by hyperhydricity
    - (*c*) the proportion of shoots which fail to become rooted and do not form plantlets and
    - (*d*) the proportion of plantlets which do not survive acclimatization
6. The final number of plants required and whether production is a single event or part of a continuing process.

28.8. METHODS TO REDUCE COST OF MICROPROPAGATION

28.8.1. Reduction of Labour Costs

The wage costs depend on the efficiency of the laboratory, the kinds of plants being grown and the methods of micropropagation being employed. As the greatest proportion of costs in propagating plants using shoot or nodal culture is involved in labour required for transfers, wage costs can be reduced by:

(*a*) *Reducing Handling Time*

- Shortening the time taken for an operator to deal with each container

It takes longer and is hence, more expensive to dissect out separate shoots than it does to cut shoots randomly, or to divide a mass of shoots into two or more pieces, or to move clumps of multiplying shoots to fresh media without division. Instead of dissecting out individual shoots to serve as explants for subcultures, it is sometimes possible to cut off shoots quickly, pass on those shoots of a sufficient size to stage III, and then chop the remaining basal shoot mass into two or more crude pieces to initiate new cultures.

- Reducing or simplifying the stages required for micropropagation

Avoiding the need for transfer by adding liquid medium containing appropriate growth regulators to the top of spent agar medium (Two-phased media additions) is particularly effective when there is no need to harvest shoots for multiplication or transfer to stage III. Roots may also be produced in this two-phased medium, or shoots can be harvested after a brief period of exposure to be rooted *ex vitro*.

- Ensuring that vessels are of such nature that they can be quickly opened and closed and that explants can be easily inserted and removed.

(*b*) *Reducing cost of Washing and Cleaning*

The cost of washing vessels can be reduced by using cheap disposable plastic containers. Vessels which has contained liquid media can go straight to the dish-washer without pre-treatment.

(*c*) *Laboratory Design*

It is always wise to try to facilitate rapid and efficient movement throughout the laboratory complex by efficient laboratory design. However, wider aisles and shelves though makes it much easier and quicker to place and retrieve cultures from growth rooms, it leads to extra non-productive space.

28.8.2. Choice of Containers

Many of the commercial laboratories prefer to place many cultures in one large box compared to smaller vessels containing only a few explants. Large plastic boxes have several advantages:

- Less vessels are required for a given number of plants
- Less effort is required to add medium and clean vessels after use
- Explants are easily inserted, cultures easily taken out
- Less vessels have to be dealt with if a liquid supplement has to be added during culture

However, lid removal can be time-consuming and requires the use of both hands, but once the vessel is opened, many cultures can be dealt with before it has to be resealed. The greatest disadvantage from using large vessels is that if there is contamination, many cultures may have to be discarded.

28.8.3. The Cost of Materials

Media costs form only a small proportion of the total cost of running a micropropagation laboratory. Costs can be reduced if agar or chemicals are ordered in bulk. However, the cost of any single ingredient in the medium is small compared to the cost of labour required for micropropagation.

The use of liquid instead of semi-solid media, not only saves the cost of adding a gelling agent, but also reduces the cost of washing and cleaning. Media costs can also be reduced by culturing more shoots per culture vessel.

The cost of electricity to power heating, cooling and lighting within a laboratory is significant. In many countries, electricity can be purchased more cheaply during the night than during the day. In these circumstances, commercial laboratories using a 12 or 16 hour photoperiod usually arrange for part of it to coincide with the 7 or 8 hour period of cheap supply. Low power lights mounted on the ceiling are turned on to gain access to rooms at other times. The possibility of reducing the number of lighting tubes in a growth room by arranging them in vertical, rather than horizontal banks is also preferred.

28.8.4. Preparation for Planting and Dispatch

Shoots rooted *in vitro* in an agar medium are particularly difficult to deal with as usually, they must be washed free of agar before planting. Considerable savings in labour costs can be made by rooting shoots in liquid medium, or even better in porous materials. Shoots rooted *in vitro* in plugs can be packed for dispatch to customers with the minimum of preparation.

One of the most effective cost-reducing steps, which have been generally followed by commercial laboratories, is to dispense with, wherever possible, the rooting of shoots *in vitro*. By transferring shoots directly from stage II to be rooted, and at the same time, acclimatized in 'soil', a whole *in vitro* cultural stage is dispensed with.

28.8.5. Increasing Plant Multiplication Rates

One important factor influencing the cost of micropropagation, is how quickly plant numbers can be increased. Rapid rates of propagation are not confined to shoot and node culture and even faster multiplication is possible from some plants using direct shoot formation or embryogenesis.

28.8.6. The Importance of Timing

Efficient and low cost working requires that growth room space be kept continuously filled, because if new cultures do not immediately replace those, which have been removed, the laboratory is losing money. It is also important to have plants ready at the exact time they are required by the customers. It is easier to avoid slack periods in a production schedule if a large variety of plants is cultured, each having a peak of demand at a different time of the year.

28.8.7. Growth Room Utilization

Cost on growth room space is based on the concept that the laboratory is working to full efficiency when shelving is :

- Fully occupied all the time, and
- Vessels are kept on growth room shelves for the minimum possible interval.

Plants cost greater if growth room space is not fully occupied. Laboratories with high cost of growth room occupancy must work with high value plants or have a large throughput of plants.

Growth room utilization can be increased if mobile racks of shelving are installed, because it is then possible to eliminate all but one passage way. To gain access to a stack of shelves, racks are pushed to one side until there is sufficient space for entry. (Plate 58)

Production cost can be reduced if more than one layer of vessels is placed on growth room benches. This can be achieved by containing vessels in light wire mesh trays, one on top of another. As light penetration into vessels is often poor when lighting sources are placed above growth room benches especially if vessels are stacked one on top of another, it is better to have provision for sideways lighting.

Although growth rooms, ideally, should be fully occupied, this is generally an unrealizable objective because there will always need to be a small reserve of space. Cultures may not be produced according to the plan, and demand is likely to fluctuate. Some laboratories seem to carry a large number of stock cultures, which, by permanently occupying bench space, reduces the effective throughput of newly propagated plants.

Assuming that the growth room space is always kept fully occupied, the cost of production can be reduced by:

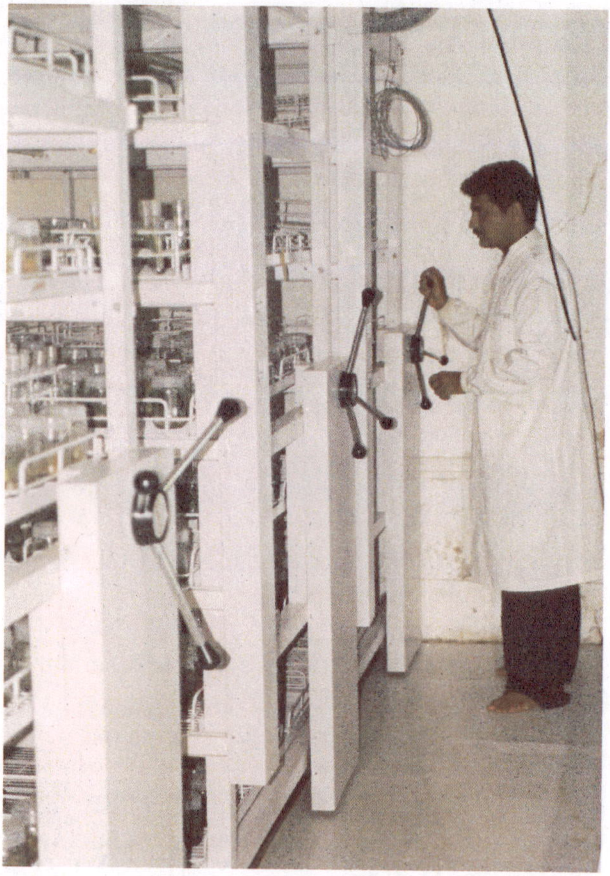

Plate 58: Mobile growth room racks in commercial units

- Placing as many explants into a vessel as is possible without causing a decrease in the rate of shoot production
- Allowing each culture the optimum amount of space
- Placing more than one layer of culture vessels upon the shelves in growth room
- Increasing the multiplication coefficient
- Reducing the time between subcultures
- Producing very close to the exact final number of shoots required in the minimum number of subcultures
- Minimizing losses due to contamination.

28.8.8. Photoautotrophic Growth of Cultures

It has been suggested that production costs could be significantly reduced by adopting autotrophic growth techniques for micropropagation. To achieve photoautotrophic growth, cultures must be subjected to light of high photon fluence. But, this means that the running costs are high if the light is provided from artificial sources. As the addition of sugar to the medium is unnecessary when cultures are photoautotrophic, problems of contamination are much reduced or eliminated. This should permit the use of large vessels, facilitating mechanical handling and automation. However, this technique seems to be applicable only to cultures at stage II and III, bearing shoots with green leaves.

28.8.9. Mechanization and Automation

Large laboratories frequently use large capacity double sided autoclaves situated between the medium preparation room and the inoculation/transfer room. This enables vessels and equipment to be loaded unsterilized on one side, and unloaded sterilized on the other.

Preparing media and dispensing it into many thousands of small vessels is a time-consuming process and one, which adds to the total cost of the present methods of micropropagation. Automatic dispensing machines can be purchased suitable for filling small containers, which use peristaltic pumps to deliver media.

Several firms sell automatic bench top sterilizers, which mix and sterilize a medium and afterwards, maintain it at a selected temperature. The medium is then passed via sterile tubing to a dispensing machine. If the dispenser is located in a laminar flow cabinet, pre-sterilized vessels can be filled without contamination.

28.8.9.1. Automation of Plant Tissue Culture Techniques

28.8.9.1.1. Use of Robots

Strategies for the automation of micropropagation being developed by several groups in Japan depend on robot-assisted cutting and selection of nodal explants containing axillary buds by computerized image analysis, followed by robotic transfer to fresh agar nutrient media. The process depends on complex and rather expensive technology, requiring a specially trained and skilled labour force. Only a few large companies are able to afford such expensive systems, which may not, in any case, necessarily reduce the costs significantly.

28.8.9.1.2. Use of Bioreactors

A completely integrated, mechanical/biological automated micro-propagation process, from the growth and multiplication to planting in soil, has been developed in Israel. It uses bioreactors to grow large and compact masses of shoot meristem. The meristems are processed, sorted and distributed automatically, and then allowed to develop into tiny rooted plantlets. The latter are transplanted automatically into soil by a planting machine. At the present time it offers a very practical method for the automation of micropropagation, with a substantial reduction in labour input and costs of production. The same technology can be readily adapted for the growth of regenerable cells and tissues in liquid media in bioreactors, as demonstrated by the successful mass propagation of shoots in a 500 litre bioreactor by Akita *et al.* (1994). They harvested 64.6 kg of shoots of *Stevia rebaudiana* from an initial inoculum of 460g, 140-fold increase. Similarly, a 10 l jar fermenter was used for the production of potato tubers, which could be used directly as seed tubers.

The most intractable problems associated with the use of bioreactors for large-scale somatic embryogenesis concern the biology of the system and engineering. Automated, bioreactor-based micropropagation may be the method of choice for the mass propagation of plants, and hopefully similar systems can be developed for the production of somatic embryos and viable synthetic seeds.

Although there is a great deal of interest in propagation using bioreactors, commercial applications have been slow to develop.

Requirements of Bioreactors

Due to the varied demands made by biological systems on their environment, there is no universal bioreactor. However, the general requirements of the bioreactor are as follows:

1. The design and construction of biochemical reactors must preclude foreign contamination (sterility). Furthermore, aseptic conditions should be maintained during the process and ensure containment.
2. Optimal mixing with low, uniform shear
3. Adequate mass transfer (of oxygen)
4. Clearly defined flow conditions
5. Feeding of substrate with prevention of under or overdosing
6. Suspension of solids
7. Gentle heat transfer
8. Compliance with design requirements, such as ability to be sterilized, simple construction, simple measuring, control and regulating techniques, scale up, flexibility, long-term stability, compatibility with up-downstream processes, and antifoaming measures.

Types of Bioreactors

Bioreactors can be classified according to various different criteria

(*a*) ***Type and form of Biocatalyst*:** This is based on free cells in submerged cultures, carried bound or immobilized cells/enzymes, and retention or re-circulation of the biocatalyst

(*b*) ***Configuration*:** Tank (height/diameter <3), Column (height/diameter > 3)
(*c*) ***Energy input and aeration*:** liquid phase, gas phase, combined.
(*d*) ***Hydrodynamics*:** perfect mixing, partial mixing, no mixing
(*e*) ***Mode of operation*:** batch, continuous, fed-batch.

28.8.9.2. Types of Bioreactors

28.8.9.2.1. Stirred Tank Bioreactors (STB)

Stirred tank reactor is the choice for many (more than 70%) though it is not the best. STBs have the following functions: homogenization, suspension of solids, dispersion of gas-liquid mixtures, aeration of liquid and heat exchange. The STB is provided with a baffle and a rotating stirrer is attached either at the top or at the bottom of the bioreactor. The typical decision variables are: type, size, location and the number of impellers, sparger size and location. These determine the hydrodynamic pattern in the reactor, which in turn influence mixing times, mass and heat transfer coefficients, shear rates, etc. The conventional fermentation is carried out in a batch mode. Since stirred tank reactors are commonly used for batch processes with slight modifications, these reactors are simple in design and easier to operate. Many of the industrial bioprocesses even today are being carried out in batch reactors though significant developments have taken place in the recent years in reactor design, the industry, still prefers stirred tanks because in case of contamination or any other sub-standard product formation the loss is minimal. The batch stirred tanks generally suffer due to their low volumetric productivity. The downtimes are quite large and unsteady state fermentation imposes stress to the microbial cultures due to nutritional limitations. The fed batch mode adopted in the recent years eliminates this limitation. The STBs offer excellent mixing and reasonably good mass transfer rates. The cost of operation is lower and the reactors can be used with a variety of microbial species. Since stirred tank reactor is commonly used in chemical industry the mixing concepts are well developed. STR with immobilized cells is not favoured generally due to attrition problems, however, by separating the zone of mixing from the zone of cell culturing one can successfully operate the system.

28.8.9.2.2. Airlift Bioreactors (ALB)

Airlift bioreactors (ALB) are generally classified as pneumatic reactors without any mechanical stirring arrangements for mixing. The turbulence caused by the fluid flow ensures adequate mixing of the liquid. The draft tube is provided in the central section of the reactor. The introduction of the fluid (air/liquid) causes upward motion and results in circulatory flow in the entire reactor. The air/liquid velocities will be low and hence the energy consumption is also low. ALBs can be used for both free and immobilized cells. There are very few reports on ALBs for metabolite production. The advantages of Airlift reactors are the elimination of attrition effects generally encountered in mechanical agitated reactors. It is ideally suited for aerobic cultures since oxygen mass transfer coefficient are quite high in comparison to stirred tank reactors.

28.8.10. Alternative Cultural Techniques

Besides improving the means by which the plants are micropropagated, it is necessary to consider whether costs could be reduced by adopting an alternative method. Plants can be grown *in vitro* in many different ways; the nature of the container can be varied, different materials can be used for support, and nutrients can be delivered by a variety of mechanisms.

28.8.10.1. Shoot Hedges

A scheme which makes use of a double-layered medium may overcome the need for regular subcultures. Dense shoots cultures could be established in such a medium and once the shoots grow to about 100 mm size, they can be cut off to serve either as new explants or as microcuttings, while the shoot bases are left to produce new axillary shoots. This method is possible only for large-scale continuous production of microcuttings.

28.8.10.2. Aeroponic Culture

Spraying callus tissue with nutrient medium, or maintaining it in a nutrient mist has been advantageous in the production of biomass or secondary products. Culture of cells on layers of a permeable support, which is then sprayed with nutrient mist, in artificial light, also has been attempted.

28.8.10.3. Direct Shoot Formation

Adventitious shoots formed directly from explants, can be used for micropropagation of plants from which this type of regeneration occurs readily. In some of these species, shoot multiplication can be carried out in an agitated liquid medium.

28.8.10.4. Encapsulated Shoot Buds

It has been found that if shoot buds are encapsulated, they produce adventitious roots and germinate to produce plants when planted in soil.

28.8.10.5. Macerated Tissues

The automated propagation of plants would be facilitated if plants, devoid of somaclonal variation, could be regenerated from tissue, which had been cut into pieces mechanically.

29

MICROSCOPIC AND PHOTOGRAPHIC TECHNIQUES FOR PLANT TISSUE CULTURE

Visual observations of plant cell and tissue culture are necessary for study of various parameters of molecular genetics, virology, and physiology because experiments based on these aspects need visual proof. For example, whether or not cultured tissue undergoes embryogenesis or organogenesis is confirmed by visualizing the presence of somatic embryos or organs.

29.1. MICROSCOPY

The microscope is an instrument designed to make fine details visible.

29.2. THE CONCEPT OF MAGNIFICATION

The image of an object can be magnified when viewed through a simple lens. By combining a number of lenses in the correct manner, a microscope can be produced that will yield very high magnification values.

29.3. LENSES AND GEOMETRICAL OPTICS

The action of a simple lens, similar to many of those used in the microscope, is governed by the principles of refraction and reflection and can be understood with the aid of a few simple rules about the geometry involved in tracing light rays through the lens.

29.4. MICROSCOPE OPTICAL COMPONENTS

The optical components contained within modern microscopes are mounted on a stable, ergonomically designed base that allows rapid exchange, precision centering, and careful alignment between those assemblies that are optically interdependent. Together, the optical and mechanical components of the microscope, including the mounted specimen on a glass micro slide and cover slip, form an **optical train** with a central axis that traverses the microscope base and stand.

29.5. MICROSCOPE ILLUMINATION

One of the most critical aspects in optical microscopy is to ensure that the specimen is illuminated with light, which is bright, glare-free, and evenly dispersed in the field of view.

29.5.1. Image Brightness

Regardless of the imaging mode utilized in optical microscopy, image brightness is governed by the light-gathering power of the objective, which is a function of numerical aperture. Just as brightness of the microscope source illumination is determined by the square of the condenser working numerical aperture, brightness of the specimen image is proportional to the square of the objective numerical aperture.

29.6. MICROSCOPE OBJECTIVES

Microscope objectives are the most important components of an optical microscope because they determine the quality of images that the microscope is capable of producing. There is a wide range of objective designs available that feature excellent optical performance and provide for the elimination of most optical aberrations.

- **Specifications and Identification:** Microscope manufacturers offer a wide range of objective designs to meet the performance needs of specialized imaging methods, to compensate for cover glass thickness variations, and to increase the effective working distance of the objective. Often, the function of a particular objective is not obvious simply by looking at the construction of the objective, but many specifications are permanently engraved on the objective barrel.
- **Objectives for Specialized Applications:** Standard brightfield objectives, corrected for varying degrees of optical aberration, are the most common and are useful for examining specimens with traditional illumination techniques. Other, more complex, methods require specific objective configurations, which often include placement of a detector on or near the rear focal plane. To complicate the issue, the objective rear focal plane can reside in the centre of an internal glass lens element, an area that is not easily accessible to the microscopist.
- **Water Immersion Objectives:** Microscopic investigations of thinly cut fixed tissue sections and living cells adhered to glass substrates routinely produce superb high-resolution images when employing plan apochromat

or fluorite objectives having a high numerical aperture. However, a significant amount of current biological research involves the investigation of cellular dynamics inside living tissue where important events can occur deep within the specimen, far away from the cover glass. Attempts to image cellular details and activities at a micrometer distances from the specimen cover glass with conventional oil immersion techniques often suffer from artefacts, including severe optical (spherical) aberration. The use of water in place of oil, as the immersion medium, is an effective approach to overcoming the aberration problems, and highly corrected **water immersion** objectives have been introduced by several manufacturers for applications involving living cells and tissues.

- **Adjustment of Objective Correction Collars:** Most microscope objectives are designed to be used with a cover glass that has a standard thickness of 0.17 millimetres and a refractive index of 1.515, which is satisfactory when the objective numerical aperture is 0.4 or less. However, when using high numerical aperture dry objectives (numerical aperture of 0.8 or greater), cover glass thickness variations of only a few micrometers result in dramatic image degradation due to aberration, which grows worse with increasing cover glass thickness. To compensate for this error, the more highly corrected objectives are equipped with a correction collar to allow adjustment of the central lens group position to coincide with fluctuations in cover glass thickness.
- **Numerical Aperture and Resolution:** Numerical aperture as applied to microscope objectives is a measure of the ability to gather light and resolve fine specimen detail at a fixed object distance. The resolution of a microscope objective is defined as the smallest distance between two points on a specimen that can still be distinguished as two separate entities. Resolution is a somewhat subjective value in microscopy because at high magnification, an image may appear unsharp but still be resolved to the maximum ability of the objective. Numerical aperture determines the resolving power of an objective, but the total resolution of a microscope system is also dependent upon the numerical aperture of the substage condenser. The higher the numerical aperture of the total system, the better the resolution.
- **Image Formation:** In the optical microscope, image formation occurs at the intermediate image plane through interference between direct light that has passed through the specimen unaltered and light diffracted by minute features present in the specimen. The image produced by an objective lens is conjugate with the specimen, meaning that each image

point at the intermediate plane is geometrically related to a corresponding point in the specimen.

- **Optical Aberrations:** Departures from the idealized conditions of Gaussian optics are known as optical aberrations. Microscope optical trains typically suffered from as many as five common aberrations: spherical, chromatic, curvature of field, comatic, and astigmatic. Geometrical distortion is another artefact often encountered in the zoom lens systems found in stereoscopic microscopes.
- **Immersion Media:** Most low-power objectives are designed to be used "dry" with air as the imaging medium. Higher magnification objectives commonly use liquid immersion media to help correct aberrations and increase numerical aperture.
- **Mechanical Tube Length:** The mechanical tube length of an optical microscope is defined as the distance from the nose-piece opening, where the objective is mounted, to the top edge of the observation tubes where the eyepieces (oculars) are inserted. Until the 1980s, most microscopes had a fixed tube length ranging from 160 to 210 mm, depending upon the manufacturer and application. Modern microscopes are equipped with infinity-corrected objectives that utilize a tube lens in the microscope body to form a parallel region of light waves into which optical accessories can be inserted without seriously affecting image quality.
- **Modulation Transfer Function:** The modulation transfer function of a lens, microscope objective, or other optical system is a measurement of its ability to transfer contrast at a particular resolution level from the object (or specimen) to the image. Computation of the modulation transfer function is a mechanism that is often utilized by optical manufacturers to incorporate resolution and contrast data into a single specification.

29.7. INFINITY OPTICAL SYSTEMS

In modern research-grade microscopes equipped with infinity-corrected optical systems, the objective no longer projects the intermediate image directly into the intermediate image plane. Instead, the objectives are designed so that light emerging from the rear aperture is focused to infinity, and a second lens, known as the **tube lens**, form the image at its focal plane.

29.8. EYEPIECES (OCULARS)

Eyepieces work in combination with microscope objectives to further magnify the intermediate image so that specimen details can be clearly observed. There are two major types of eyepieces that are grouped according to lens and aperture diaphragm arrangement: the negative eyepieces with an internal diaphragm and positive eyepieces that have a diaphragm below the lenses of the eyepiece. In many instances, eyepieces are designed to work together with objectives to eliminate chromatic aberration.

29.9. SUBSTAGE CONDENSERS

The substage condenser gathers light from the microscope light source and concentrates it into a cone of light that illuminates the specimen with uniform intensity over the entire viewfield. It is critical that the condenser light cone be properly adjusted to optimize the intensity and angle of light entering the objective front lens. Perhaps the most poorly understood component in the optical train, the condenser is nevertheless one of the most important factors in obtaining high quality images in the microscope.

29.10. SPECIMEN STAGES

All microscopes are designed to include a stage where the specimen (usually mounted onto a glass slide) is placed for observation. Stages are often equipped with a mechanical device that holds the specimen slide in place and can smoothly translate the slide back and forth as well as from side to side. Other stages are designed to allow rotation of the specimen through 360 degrees or to provide anchors for auxiliary light sources, specimen manipulation tools, and other accessories.

29.11. REFLECTED LIGHT MICROSCOPY

Microscopy using oblique or epi-illumination is utilized for the study of specimens that are opaque, including semiconductors, ceramics, metals, polymers, and many others.

29.12. FUNDAMENTALS OF STEREOMICROSCOPY

Considering the wide range of accessories currently available for stereomicroscope systems, this class of microscopes is extremely useful in a multitude of applications. Stands and illuminating bases for a variety of contrast enhancement techniques are available from all of the manufacturers, and can be adapted to virtually any working situation. There are a wide choice of objectives and eyepieces, enhanced with attachment lenses and coaxial illuminators that are fitted to the microscope as an intermediate tube. Working distances can range from 3-5 cm to as much as 20 cm in some models, allowing for a considerable amount of working room between the objective and specimen.

29.13. PHOTOGRAPHIC TECHNIQUES

Various photographic techniques are required in order to document various aspects of plant cell and tissue culture, because objects of interest vary greatly in size and type, but are often very small and hence, equipment to accomplish both macro- and microphotography are required. The use of photography to capture images in a microscope dates back to the invention of the photographic process. Early photomicrographs were remarkable for their quality, but the techniques were laborious and burdened with long exposures and a difficult process for developing emulsion plates. The primary medium for photomicrography was film until the past decade when improvements in electronic camera and computer technology made digital imaging cheaper and easier to use than conventional photography. This section will address photomicrography both on film and with electronic analog and digital imaging systems.

The quality of a photomicrograph, either digital or recorded on film, is dependent upon the quality of the microscopy. Film is a stern judge of how good the microscopy has been prior to capturing the image. It is essential that the microscope be configured using Köhler illumination, and that the field and condenser diaphragms are adjusted correctly and the condenser height is optimized. When properly adjusted, the microscope will yield images that have even illumination over the entire field of view and display the best compromise of contrast and resolution.

Almost all microscopists will, at some point, have the need or desire to record the images seen through the microscope. The main mechanism, for many years, of producing such photomicrographs was through the use of film, although in recent years most scientists have begun to capture images by means of electronic cameras.

29.14. TYPE OF PHOTOGRAPHY FOR PLANT CELL AND TISSUE CULTURE RESEARCH

29.14.1. Macrophotography

When cultures contained in whole Petri dish or groups of Petri dishes must be photographed, a simple 35mm SLR camera equipped with a close-up lens and mounted on a copy stand is preferable. In this instance, the photographic field might range from 2cm (a large callus culture) to 10 cm (a Petridish) to 40 cm (a group of Petridishes).

29.14.2. Microphotography

Stereomicrography: Stereomicrography is suitable for the characterization of callus cultures, small shoots, nonzygotic embryos, and other objects that range in size from 0.5 mm to 1.5 cm. A stereomicroscope is used for examining the surface morphology of a specimen. As specimens often possess much surface relief, the ability to photograph a wide depth of field is essential in stereomicrography. Depth of field is usually adjusted on those stereomicroscopes that incorporate an iris diaphragm between the objective lens and the eyepieces. It increases as the iris diaphragm is closed. Stereomicroscopes often use a separate light source. The light source for stereomicrography consists of a fibre optic illuminator with a bifurcated fibre bundle. This facility allows great flexibility in lighting with an optimum view of the specimen.

***Compound Micrography*:** Compound micrography is suitable for small objects (early stage embryos or cells, or the internal structure of cells and tissues). This can resolve objects from the 10µm to 1.5 mm size range, depending on optical quality. A compound microscope utilizes light that is transmitted through the specimen. The specimen must be thin and lucent, which can allow enough light to pass through. Typical histological processing of tissue is employed to obtain lucent sections.

Compound microscopes are of two types: Upright and Inverted. **Upright microscopes** are the most common and typically have the highest resolving power. They are usually used to view specimens mounted on microscope slides with cover slips (the specimen is below the objective lens). Inverted microscopes are used to examine cells and tissues in liquid, such as suspension cultures or protoplasts. Here specimen is viewed through the bottom of the culture vessel (specimen is placed above the objective lens). **Inverted microscopes** are indispensable in protoplast research, where living cultures must be monitored *in situ*.

***Fluorescence Microscopy*:** Fluorescence illumination and observation is the most rapidly expanding microscopy technique employed today, both in the medical and biological sciences, a fact which has spurred the development of more sophisticated microscopes and numerous fluorescence accessories. Epi-fluorescence, or incident light fluorescence, has now become the method of choice in many applications. In contrast to other modes of optical microscopy that are based on macroscopic specimen features, such as phase gradients, light absorption, and birefringence, fluorescence microscopy is capable of imaging the distribution of a single molecular species based solely on the properties of fluorescence emission. Thus, using fluorescence microscopy, the precise location of intracellular components labelled with specific fluorophores can be monitored, as well as their associated diffusion coefficients, transport characteristics, and interactions with other biomolecules. (Plates 59 and 60)

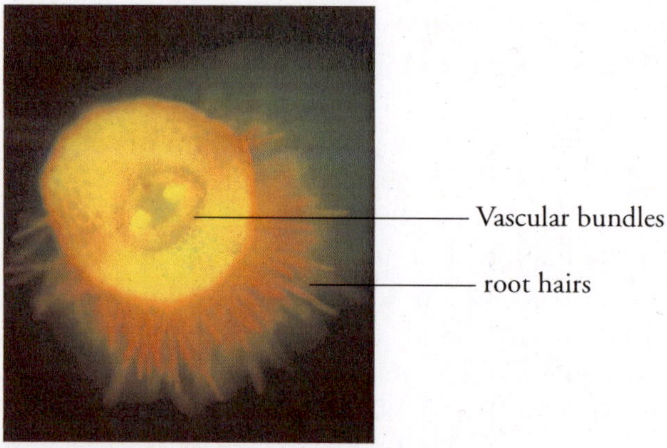

Plate 59: Cross section of root showing root hairs and vascular bundles

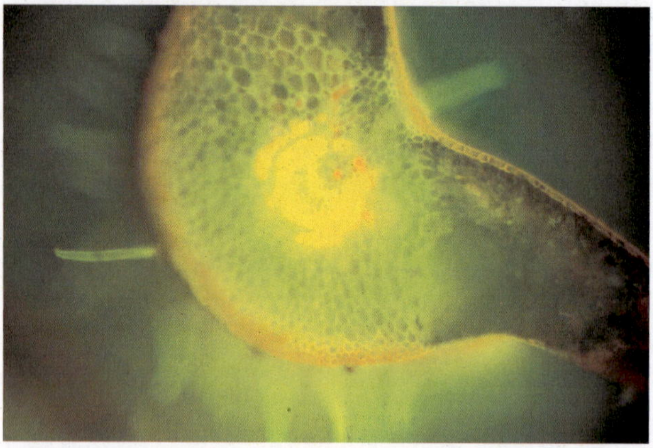

Plate 60: Cross section of leaf showing midrib with vascular bundles inside

Certain biological molecules fluoresce when excited with an intense beam of light of a particular wavelength. In histological specimens, fluorescence may be excited in compounds inherent in the tissue (autofluorescence). The range of plant tissues that contain fluorescent molecules is limited. Hence, this technique is extended with the use of fluorochromes. The fluorochromes are compounds, which bind to some specific chemical molecules in the tissues and fluoresce when excited with a particular wavelength of light.

Several fluorochromes are available for a range of tissues. Acridine Orange, Calcofluor and Auramine 'O' are the commonly used ones. Acridine Orange is generally known to stain RNA and DNA and is particularly used to reveal nuclei. With Acridine Orange, nuclei fluoresce orange when excited under blue light. Calcofluor is a useful fluorochrome, which facilitates detailed study of changes occurring in tissues as it binds to cell walls. Cells and tissues stained with

Calcofluor fluoresce blue when excited under UV light. Auramine 'O' binds to two important chemical constituents of lignin and cutin and fluoresce yellow when excited under blue light. Auramine 'O' when used in combination (double stained) with Calcofluor provides more contrasting clarity of different tissue of a plant specimen.

Preparing plant material for fluorescence microscopy observations: Preparing a plant material for observation under fluorescence microscope is probably the easiest and is quicker. In a day, an interested and dedicated student can probably make a couple of hundreds of sections. A simple procedure could be as follows:

29.15. PROCEDURE FOR JACKFRUIT PLANT SAMPLE PREPARATION

29.15.1. Materials Required

- Conventional hand-held microtome.
- A pith material, or alternatively fresh tender carrot roots
- Petri plates, shaving razor blades of fine quality
- Alcohol, fluorochromes or stams
- Razor scalpels

29.15.2. Procedure

Microshoots which were placed in the rooting media are removed from the culture tubes, base of the shoots is rinsed in water to remove agar adhering to it.

The base of the microshoot which was submerged in the media was cut. This shoot is inserted into a slit made in the pith. The pith is inserted into the hand-held microtome of properly fastened. Using a razor blade, sections of pith is sliced along the cut surface of the microshoots. As many cross sections are taken. It is good to take thin sections but not a must if an inverted microscope is used (Inverted microscope facilitates observation of just the top surface of the section).

In cases where hand-held microtomes are not available carrots could be used instead of pith. Microshoots are placed on slit made on carrots and carrot could be held in hand and top in sliced in such a way that a thin section in obtained from base of the microshoot.

The sliced sections of the shoot (either cross section or longitudinal section) is placed in a Petri dish containing 70% alcohol. This would arrest any further changes which may take place in the cut sections. After a couple of minutes these sections are transfered to fluorochrome solutions (0.01%) for approximately 3-5 minutes. After rinsing in water to remove excess stain, the sections are mounted on glass sides in double distilled water. Alternatively, the sections are transferred to water in Petri dish or alternatively in replidishes when sections are required to be stored for prolonged periods of time.

29.15.3. Light Sources

In order to generate enough excitation light intensity to furnish secondary fluorescence emission capable of detection, powerful light sources are needed. These are usually either mercury or xenon arc (burner) lamps, which produce high-intensity illumination powerful enough to image faintly visible fluorescence specimens. A key feature of fluorescence microscopy is its ability to detect fluorescent objects that are sometimes faintly visible or even very bright relative to the dark (often black) background. In order to optimize this feature, image brightness and resolution must be maximized.

29.16. TYPES OF FILMS

Film is the recording medium for photography. Both colour positive (slide) and black and white negative (B & W) film types in 35 mm format can be used.

29.16.1. Colour Slide Film

It is used to obtain life-like, detailed images. 35 mm slides can be conveniently stored. These are useful for oral presentation of results. The film is highly sensitive to colour balance of the light used for illumination. For illumination, tungsten balanced compatible colour film should be used. The light source should be a copy stand, fibre optic illuminator, or built into the microscope. It must be adjusted to the maximum or near-maximum level. This will help to achieve the proper colour balance and is compatible with tungsten film.

29.16.2. Black and White Film

It provides a negative that can be used to produce black and white prints. They are most commonly used for presenting results in a publication. It has no colour balance problems. The quality of prints obtained from colour negatives differ tremendously with the film type.

29.17. COMPOSITION

The composition of the subject in the viewing area often makes the difference between a poor photograph and an excellent photograph. The specimen should be carefully cropped (to fill most of the viewing area). Also, the orientation and lighting of the specimen are critical.

30
HISTOLOGICAL TECHNIQUES

Plant histology can be defined as the study of the microscopic structures or characteristics of cells and their assembly and/or arrangement into tissues and organs. The techniques used include those for brightfield and fluorescent microscopy, in which the specimens can be prepared for cutting into thick sections (15 to 40 μm) either without a stabilizing medium (fresh sections), in cryofluids (frozen sections), or embedded in paraffin-like materials, or in various formulations of plastic. Other techniques that employ electron microscopy either do not require special embedding medium for specimen preparation (scanning electron microscopy) or have samples embedded in plastics (transmission electron microscopy) that can be cut into ultra-thin sections (65 to 100 nm).

Histological development can be studied overtime by periodically sampling tissues and/or the results can be observed in the mature structure. Histological sections are static and fixed in time. Nevertheless, the origin of tissues and organs may be convincingly inferred from serial observations of many samples and sections.

30.1. GENERAL CONSIDERATIONS

30.1.1. Equipment and Supplies

- Plastic beakers, aluminum pans
- Rotary microtome with disposable blade holder
- Warming oven, trays
- Waterbath
- Hot plate
- Disposable plastic base molds and embedding rings
- Screw capped coplin jars
- Clean glass slides and cover slips
- Camel hair brush
- Isopropanol, Alcian blue, Histochoice fixative, ethanol, Paraplast embedding medium, Micro-clear

30.2. GENERAL PROTOCOL FOR PARAFFIN STUDIES

30.2.1. Fixation

- Identify typical specimens (e.g., embryos on leaf sections), remove them from culture dishes, and place in a small Petri dish containing water.
- Carefully trim the specimen using a blade or scalpel to include the subject and a small area of the surrounding tissue.
- Place a few specimens in about 5 ml of Histochoice, a nonaldehyde, non-toxic fixative amended with 20% ethanol in specimen vials. (It is advisable to aspirate the open vials to remove air from the samples so that it promotes better infiltration of the fixative).
- Close the vials and allow the tissue to remain in the fixative at room temperature for a minimum of 24 hours. Samples may be stored at room temperature in the fixative for weeks or months without degeneration of structures.

30.2.2. Dehydration and Infiltration

- After fixation, dehydrate the sample and place in a solvent compatible with paraffin. (Traditionally, this has been accomplished with graded series of ethanol and transitioning to tertiary butyl alcohol. This process can be considerably shortened by using isopropanol to both dehydrate the tissue and dissolve the paraffin).
- Remove the fixative from the vials with a pipette and dehydrate the specimens using 30, 50, 75, 95 and three changes of 100% isopropanol, each for about 30 minutes.
- After the last change of isopropanol, fill the vials about one-quarter full with fresh 100% isopropanol and add a few pellets of Paraplast embedding medium, loosely recap the vials and incubate at 60°C. periodically over the next several hours, swirl the contents and add a few more pellets of Paraplast to each of the vials. At the end of the day, remove the caps from the vials to allow the isopropanol to evaporate. Fill a plastic beaker with Paraplast pellets, place several pellets into a number of base molds contained in an aluminium pan and incubate at 60°C in an oven overnight. By the next day, all the isopropanol would have dissipated, the specimens completely infiltrated with Paraplast, and the pellets in the beaker and molds would have melted.

30.2.3. Casting Specimens into Blocks

- Transfer the specimen into molten paraffin contained in a base mold and position it so that the sectioning plane of the specimen is parallel to the bottom of the base mold.

- Once the specimen is oriented correctly, quickly transfer the base mold to a cool surface (like an inverted pan containing ice). The paraffin surrounding the base mold will quickly become congealed and specimen will be immobilized. Rapidly attach an embedding ring to the base mold and fill the apparatus with molten paraffin from the plastic beaker (If the immobilizing paraffin is allowed to congeal too far, the added molten paraffin will not bond with it and the block will split apart).
- Float the casted blocks in a large beaker containing cold water.
- Remove the base mold after the block is hardened.

30.2.4. Microtomy and Mounting Sections on Slide

- Trim away some of the excess paraffin from around the sample.
- The horizontal edges of the block must be parallel to each other. The block edges must also be kept parallel to the knife edge.
- Set the microtome to cut sections of the desired thickness, say 10 mm. Mount the specimen block in the chuck and orient it so that the bottom of the cutting face is parallel to the edge of the blade.
- Position the specimen at the level of the knife and carefully and gently slide the knife holder towards the block until they are very close.
- Turn the wheel clockwise with care causing the specimen to advance 10 mm per revolution. Continue to operate the microtome and after five or more sections are obtained, place a camel hairbrush underneath the ribbon for support while it is held away from the knife holder.
- Lay the ribbon out on a clean piece of dark paper and using a blade, divide it into segments that do not exceed about 75% the length of the slide.
- Moisten the tip of a brush, touch it to one of the section and lift it off the paper. Now touch one end to a drop of water on a slide to lay the ribbon down. Take care to see that no air is trapped underneath the sections.
- Warm the slide at 50°C to expand the sections. Affix the sections to the glass by placing the slide on a warming table at 40°C overnight. The sections are now ready to be stained.

30.2.5. Staining of Sections (with Alcian Blue and Hematoxylin)

- Preheat the alcian blue solution in a closed coplin jar and place the slides in the stain for 1h at 50°C.
- Gently rinse the slides in a stream of cold tap water and carefully blot dry the remaining water droplets.

- Transfer the slides to coplin jars containing hematoxylin stain for 15 minutes at room temperature, rinse and dry and place on a warming table at 40°C for a minimum of 30 minutes.
- Remove the paraffin in the sections with three 5 minute soaks in Micro-clear followed by three 5 minute changes of 100% isopropanol.
- After the slides have been air dried, pool some resin in the centre of the slide and gently lower a coverglass so that the resin spreads evently over the surface without air bubbles. Cure the resin overnight on a 40°C warming table.

GLOSSARY

Adaptation: The process of change in structure and/or function that makes an organism or a population better suited to survive in an environment. Adaptation may be achieved by phenotypic tuning to prevailing environmental conditions, or through evolutionary changes of genetic structure at the population level.

Adventitious: Produced at an abnormal or unusual position, or at an unusual time of development. Developing from unusual points of origin, such as shoots or root tissues from callus or embryos from sources other than zygotes. In tissue cultures, roots developing from the microshoots, shoots, which might develop from roots, are termed as adventitious in nature.

Adventitious roots: Any root, which grows from a tissue other than the pericycle or endodermis of an older root.

Agar: A gelatinous polysaccharide obtained from the red alga *Gelidium corneum* and from several other red algae. It is a solidifying agent that when mixed with nutrient media (0.6-1.0%) forms a gel for growing tissue cultured plants and for other purposes. Agar gels melt at about 100°C but solidify at about 44°C.

Androgenesis: Development of haploid plants from the male gametophyte following a developmental pattern resembling embryogenesis, resulting from the culture of anthers or microspores.

Anther culture: Refers to the culture of anthers containing the male gametophytes or microspores with the objective of producing monohaploid plants.

Artificial seed: Encapsulated or coated somatic embryos (embryoids) that are planted and treated like seeds.

Asepsis: Without infection or contaminating microorganisms.

Aseptic Technique: Procedures used to prevent the introduction of fungi, bacteria, viruses, mycoplasma or other microorganisms into cultures.

Autoclave: Apparatus in which media, glassware, etc. are sterilized by steam under pressure.

Auxins: A class of plant hormones, which induces cell elongation; often induces adventitious roots, and inhibits adventitious shoots. Endogenous auxins are auxins that occur naturally. Indole-3-acetic (IAA) is a naturally occurring auxin. Exogenous auxins are auxins that are man-made or synthetic. Examples of exogenous auxins included 2,4-Dichloro-phenoxyacetic acid (2,4-D), Indole-3-Butyric acid (IBA), Naphthaleneacetic acid (NAA), and 4-Chlorophenoxyacetic acid (CPA).

Auxotroph: A microorganism or plant cell line possessing atypical nutritional requirements for some growth factors in addition to those required by the control or wild type which can synthesize them.

Axenic culture: A culture without foreign or undesired life forms. An axenic culture may include the purposeful co-cultivation of different types of cells, tissues or organisms.

Axillary buds or branches: Shoot buds formed at the juncture of the leaf and the stem (the axil).

Batch culture: A suspension culture in which cells grow in a finite volume of nutrient medium and follow a sigmoidal pattern of growth between subcultures.

Biodiversity: The variability among living organisms from all sources including, *inter alia*, terrestrial, marine and other aquatic ecosystems and the ecological complexes of which they are part; this includes diversity within species, between species and of ecosystems.

Bioreactor: A vessel such as a fermenter, used for fermentation or other biological processes.

Biotechnology: Any technological application that uses biological systems, living organisms, or derivatives thereof, to make or modify products or processes for specific use.

Biotransformation or Biochemical transformation: The use of cultured cells to convert substrates into desired organic compounds by virtue of an endogenous enzyme which catalyses the reaction.

Buffer: A substance in solution, or system that resists pH change despite the addition of small amounts of acid or base.

Callus: Disorganized tumor-like masses of plant cells that form in culture. The process of callus formation is called **callogenesis**.

Cell: The smallest structural unit of an organism that is able to grow and reproduce independently.

Cell culture: Growing of cells *in vitro* including the culture of single cells or small aggregates of cells in a liquid medium.

Chemically Defined Medium: A nutritive solution for culturing cells in which each component is specifiable and ideally of known chemical structure.

Chimera: A plant or tissue composed of more than one kind of genetic tissue. This can happen due to mutations in a cell of a very young plant.

Clone: (*a*) A group of cells, tissues, or plants, which are in principle genetically identical.

(*b*) A group of plants derived from a single individual by asexual reproduction and thus having identical genetic constitution.

Clonal Propagation: Asexual reproduction of plants that are considered to be genetically uniform and originated from a single individual or explant.

Co-cultivation: In transformation experiments, the incubation of host cells or tissues with *Agrobacterium* or another organism acting as a vehicle for a genetic vector.

Contamination: Growth of microorganisms along side of the plant tissue *in vitro*.

Continuous culture: A suspension culture held at constant volume and continuously supplied with nutrients by the inflow of fresh medium.

Culture Room: Room for incubating cultures with controlled light, temperature, and humidity.

Cybrid: A cytoplasmically hybrid cell with organelles from both parental sources, achieved by fusion of an anucleated protoplast with a normal protoplast. During subsequent cell divisions, random or non-random segregation of organelles may occur, resulting in variable parental combinations in cybrid cell lines.

Cytokinins: A class of plant hormones, which induce cell division and often adventitious buds (shoots), while inhibiting rooting. Endogenous cytokinins, cytokinins that occur naturally, include zeatin and 6-dimethylallylaminopurine (2iP). Exogenous cytokinins, cytokinins that are man-made or synthetic, include 6-furfurylaminopurine (kinetin) and 6-benzylaminopurine (BA or BAP).

De-differentiation: A process whereby specialized, non-dividing cells begin to proliferate by mitotic divisions, presumed to involve regression to an undifferentiated state.

Differentiation: The development from callus of readily recognized organs such as roots, shoots, embryos.

Embryogenesis, Embryogeny: Initiation and development of embryos from zygotes, or of bipolar structures from somatic cells that parallel the developmental path of zygotic embryos. In latter case, such embryos may be either adventitious or *de novo* in origin.

Embryoid: Embryo-like structure that is produced from a somatic cell *in vitro*.

Embryo rescue: Application of embryo culture techniques to obtain interspecific hybrid plants prior to abortion of the embryos due to post fertilization hybridization barriers.

Explant: The tissue taken from a plant or seed and transferred to a culture medium to establish a tissue culture system to regenerate a plant.

***Ex situ* conservation:** Any conservation method that entails removal of individual plants or propagating material (seed, pollen, tissue) from its site of natural occurrence, i.e., conservation "off-site" in gene banks as seed, tissue or pollen; in plantations; or in other live collections, such as *ex situ* conservation stands.

Explant: Plant part used to initiate a tissue culture, e.g., embryo, shoot meristem, microspores, leaf, flower petals, etc.

Flaming: Sterilization of instruments (forceps, scalpel) by an open flame after dipping in alcohol.

Gene: The unit of heredity transmitted from generation to generation during sexual or asexual reproduction. More generally, the term is used in relation to the transmission and inheritance of particular identifiable traits. The simplest gene consists of a segment of nucleic acid that encodes an individual protein or RNA.

Gene marker: A gene or group of genes used to identify an individual or a cell that carries it, or as a probe to mark a nucleus, chromosome, or locus. Genetic markers can be classified as biochemical markers (e.g. isozymes) and molecular markers (e.g. DNA markers).

Genetic diversity: The genetic variability within a population or a species. This is the foundation of all diversity, and loss of genetic diversity within species is increasingly recognized as an important and largely undocumented problem, at least as serious as loss of entire species. Genetic diversity can be assessed at three levels: (1) diversity within breeding populations; (2) diversity between breeding populations within any one geographic area; and (3) diversity within the species.

Genetic drift: Change in allele frequency from one generation to another within a population, due to the sampling of finite numbers of genes that is inevitable in all finite-sized populations. The smaller the population, the greater the genetic drift, with the result that some alleles are lost, and genetic diversity is reduced. Thus minimization of genetic drift is an important consideration for conservation of genetic resources.

Genetic engineering: (*a*) Genetic manipulations, by which an individual, having a new combination of inherited properties, is established.

All modern molecular biology techniques used to isolate, manipulate and transfer genes from one organism to another. One such method is using recombinant DNA.

(*b*) The directed genetic modification of an individual organism to have a new heritable trait by splicing a specific gene into the individual's geonomic DNA sequence.

(*c*) An all-inclusive term to cover all laboratory or industrial techniques used to alter the genetic machinery of organisms so that they can subsequently synthesize increased yields of compounds already in their repertoire, or form entirely new compounds, adapt to drastically changed environments, etc. Often, the techniques involve manipulating genes in ways that bypass normal sexual or asexual transmission.

Genetic resource conservation: The collection, maintenance, storage and sustainable management of genetic resources aimed at ensuring their continued existence, evolution and availability for future generations.

Genome: All the genes of a living organism. Its complete set of chromosomes (DNA), with their associated genes. Its size is generally given as its total number of base pairs.

Germplasm: The genetic material which forms the physical basis of heredity and which is transmitted from one generation to the next by means of the germ cells. Often synonymous with genetic material, when applied to plants it is the name given to seed or other material from which plants are propagated.

Gibberellins: A plant growth regulator that influences cell enlargement. Endogenous growth forms of gibberellin include Gibberellic Acid (GA_3).

Gynogenesis: Development of haploid plants from the female gametophyte by the culture of unfertilized ovules or ovaries.

Habituation: Ability of cells to grow in the absence of normally required and exogenously supplied growth factors as a result of shifts in metabolism, especially as in the case where cultured plant cells become autonomous for the production of hormone(s).

Haploid induction: Stimulation of plant formation from male or female gametes without sexual fertilization.

Hardening off: Acclimatization of *in vitro* derived plants to ambient conditions.

***In situ* conservation:** The conservation of ecosystems and natural habitats and the maintenance and recovery of viable populations of species in their natural surroundings and, in the case of domesticated or cultivated species, in the surroundings where they have developed their distinctive properties.

Internode: The portion between two nodes in a stem.

***In vitro*:** To be grown in glass (Latin). Propagation of plants in a controlled, artificial environment using plastic or glass culture vessels, aseptic techniques, and a defined growing medium.

Laminar Flow Cabinet: A work space where sterile air moves to allow the introduction of explants aseptically.

Macronutrients: Minerals needed in large amounts for normal plant development (N, P, K, Ca, Mg).

Medium: A nutritive solution, solid or liquid, for culturing cells.

Meristem: A localized group of actively dividing cells, from which permanent tissue systems i.e., root, shoot, leaf and flower are derived.

Micronutrients: Minerals needed in relatively small amounts for normal plant development (Fe, B, Zn, Cu, Mo, Mn).

Micropropagation/*In vitro* propagation: (*a*) The *in vitro* vegetative propagation of plants producing plantlets, micropropagules, or somatic embryos.

(*b*) Vegetative propagation of plants by *in vitro* technology producing plantlets, micropropagules, or somatic embryos.

(*c*) Production of plants from very small plant parts, tissues, or cells grown aseptically in a test tube or other container where the environment and nutrition can be rigidly controlled.

(*d*) Clonal propagation by *in vitro* technology, *via* organogenesis and somatic embryogenesis.

(*e*) The propagation of plants in a controlled, artificial environment, using plastic or glass culture vessels, aseptic techniques and a defined growing medium. The procedures used for *in vitro* propagation comprise: (1) the selection of suitable explants, their sterilization and transfer to nutrient medium, (2) the proliferation of shoots on multiplication medium and (3) transfer of shoots to a rooting (or storage) medium and planting out.

Microspore: A small spore produced in a microsporangium in heterosporous plants. The microspore develops into the male gametophyte. In angiosperms, the microspore is the pollen.

Morphogenesis: (*a*) The ontological development of differentiated structures.

(*b*) The development of a structure from an unorganized state to a differentiated and organized state.

Node: A point on the plant stem from which a leaf, shoot or flower originates.

Nutrient Medium: Food for the explant, which allows growth. Contains macro- and micronutrients, vitamins, sugar, and plant growth regulators.

Organ: A group of cells or tissues forming part of an organism, with a special function, e.g. a leaf, a stamen.

Organ culture: The maintenance of growth of organ primodia or whole parts of an organ *in vitro* in a way that allows differentiation and/or preservation of its structure and/or function.

Organogenesis: The process of initiation and development of a structure which shows natural organ form and/or function, the initiation of which is temporarily separated from the initiation of other organs. Such organs may be adventive or *de novo* in origin.

Passage: The transfer or transplantation of cells or tissues with or without dilution or division, form one culture vessel to another.

Passage Number: The number of times the cells or tissues in culture have been subcultured or passed.

Petiole: A leaf stalk; the portion of the plant that attaches the leaf blade to the node of the stem.

Petri dish: A flat dish made of glass or plastic with sides turned up and an opposing cover plate.

Protoplasm: The general name for all the substances and bodies inside a cell. All living organisms are made of protoplasm.

Regeneration: In plant cultures, a morphogenetic response to a stimulus that results in the products of organs, embryos, or whole plants.

Shoot Apical Meristem: Undifferentiated tissue, located within the shoot tip, generally appearing as a shiny dome-like structure, distal to the youngest leaf primordium and measuring less than 0.1 mm in length when excised.

Solid media: Nutrient media solidified with agar.

Somaclonal variation: Genetic changes, which occur during extended periods of tissue culture. It is a phenotypic variation, either genetic or epigenetic in origin, displayed among somaclones.

Sterile: Medium or object with no viable microorganism.

Subculture: Transplanting *in vitro* grown tissues to additional plates of fresh media.

Thermotherapy: Exposure of plants to elevated temperatures as a treatment for virus or mycoplasma elimination. Thermotherapy is used alone or in conjunction with meristem culture for virus elimination.

Tissue: A group of cells, of similar shape and size, which all have the same function. Plant organs usually have several kinds of tissue, e.g. leaves have epidermal, mesophyll and vascular tissue.

Tissue culture: (*a*) The growth of isolated plant cells or small pieces of tissue under controlled conditions in a sterile growth medium. The medium is designed to meet the requirement of the tissue involved, e.g., by application of hormones and nutrients.

(*b*) A general term for aseptic cell, tissue, organ and protoplast culture.

(*c*) A technique for cultivation cells, tissues, or organs of plants in a sterile, synthetic medium; includes the tissues excised from a plant and the culture of pollen or seeds.

Totipotency: A cell characteristic in which the potential for forming all the cell types in the adult organism is retained.

Tween 20: The brand name of a wetting agent or surfactant which breaks the surface tension of tissues. Commonly added to disinfecting solutions to make them more effective.

Variant: A culture exhibiting a stable phenotypic change whether genetic or epigenetic in origin.

Wilt: Collapse of foliage from water deficiency; may be temporary or permanent.

Zeatin: A natural cytokinin first isolated from corn. It is sometimes included in plant tissue culture media. It dissolves in acid.

APPENDIX

CONVERTING MOLAR SOLUTIONS TO MILLIGRAMS/LITRE AND MILLIGRAMS/LITRE TO MOLAR SOLUTIONS

1. To convert Molar solutions to milligrams per litre

1M solution = *Molecular weight of the compound in grams in one* Litre of the solution

For example: 1M solution of Kinetin = 215.2 g/L

1mM solution of Kinetin = 215.2 mg/L

2. To convert milligrams per litre to Molar solutions

$$1\text{mg/L solution} = \frac{1}{1000 \times \text{Molecular weight of the compound in gram}} \text{M}$$

$$\text{For example: 10 mg/L of IBA} = \frac{10}{1000 \times 20302} \text{M}$$

SELECTED BIBLIOGRAPHY

Akita M., Shigeoka T., Koizuimi M. (1994). Mass propagation of shoots of *Stevia rebaudiana* in a large-scale bioreactor. *Plant Cell Rep.,* **13:** 180-183.

Al Thobani and Satyanarayana B.N. (1998). Unpublished data from thesis entitled "*In-vitro* regeneration studies in Carnation." Thesis submitted to University of Agricultural Sciences, Bangalore.

Bourgin J.P. and Intsch J.P. (1967). Production fo haploid *Nicotiana* from excised stamens. *Ann. Physiol. Veg.*, 9:377-382.

Chu C.C. (1978). The N6 medium and its applications to anther culture of cereal crops. In: *Proc. Symp. Plant Tissue Culture*, Science Press, Beijing, pp. 43-50.

Dalia Mathews and Sathyanarayana B.N. (2003). Studies on *in vitro* regeneration and mutagenesis in *Bacopa monnieri* (L.) Pennell. Thesis submited to University of Agricultural Sciences, Bangalore.

Dalia Mathews and Sathyanarayana B.N. (2003). Application of Tissue culture in Horticulture crops. *State level training on Hi-tech Horticulture for staff of water shed Dept of Karnataka.* Lalbagh, State Dept. of Horticulture, August, 2003.

Debergh P.C. and Maene L.J. (1981). A scheme for commercial propagation of ornamental plants by tissue culture. *Scientia Hort.*, **14**: 335-345.

Deepu Mathew and Satyanarayana B.N. (2002). Unpublished data from thesis entitled "Studies on callus induction and regeneration *in vitro* in chillies Cv. Byadagi Dabba", Submitted to University of Agricultural Sciences, Bangalore.

Deepu Mathews and Sathyanarayana B.N. (2002). Studies on callus induction and regeneration *in vitro* in chillies Cv. Byadagi Dabba. *Proceedings of National Symposium on Plant Physiology held at University of Agricultural Sciences, Dharwad.*

Donnelly D.J., Vidaver W.E. and Lee K.Y. (1985). The anatomy of tissue cultured raspberry prior to and after transfer to soil. *Plant Cell Tissue and Organ Culture,* **4**: 43-50.

Evans D.A., Sharp W.R. and Medina-Filho H.P. (1984). Somaclonal and gametoclonal variation. *Am. J. Bot.*, 71:749-774.

Endale Gebre and Sathyanarayana B.N. (2001). TAPIOCA-A new and cheaper alternative to agar for direct *in vitro* shoot regeneration and microtuber production from nodal culture of potato. African crop Science Journal, **9(1)**: 1-8.

Gamborg O.L., and Phillips G.C. (1995). Plant Cell, Tissue and Organ Culture, Narosa Publishing House, New Delhi, pp. 348.

Gamborg, O.L., Miller, R.A. and Ojima, K., 1968, Nutrient requirements of suspension cultures of soybean root cells. *Exp. Cell Res.*, **50**: 151-158.

George K.L. and Falkinham J.O. (1986). Selective medium for the isolation and enumeration of *Mycobacterium avium-intracellulare* and *M. scrofulaceum. Can. J. Microbiology,* **32:** 10-14.

Gopalaswamy, Nuthan, and Sathyanarayana B.N. (2005). Studies on *in vitro* induction of triploidy in *Morus alba*. Thesis submitted to University of Agricultural Sciences, Bangalore.

Grout B.W.W. (1988). Photosynthesis of regenerated plantlets *in vitro*, and the stresses of transplanting. *Acta. Hort.*, **230**: 129-135.

Grout B.W.W. and Aston M.J. (1978). Modified leaf anatomy of cauliflower plantlets regenerated from meristem culture. *Ann. Bot.*, **42**: 993-995.

Guha J. and Maheshwari S.C. (1964). *In vitro* production of embryos from anthers of Datura. Nature **204:** 497.

Hartmann H.T. and Kester D.E., 1983, *Plant Propagation: Principles and Practices.* 4th Edition, Prentice-Hall, London.

Kalpana Sachdev, Prakash K.S., Prasad T.G. and Sathyanarayana B.N. (2002). Somatic embryogenesis and plantlet regeneration from node and leaf callus cultures of *Gloriosa superba* L. Journal of Plant Biology **29**(2).

Kathireshan C., Srinivasappa K.N., Farooqui A.A. and Sathyanarayana B.N. (2003). Studies on *In vitro* propagation of Coleus forskholii-A potential medicinal plant. Proceedings of global summit on medicinal plants held on September 25-30 at Mauritius. pp. 123-126.

Kedisso E.G. and Sathyanarayana B.N. (1998). Unpublished data from thesis entitled "Studies on efficiency of *in vitro* shoot proliferation and microproduction in potato (*Solanum tuberosum* L.)", submitted to University of Agricultural Sciences, Bangalore.

Kedisso E.G. (1998). Studies on efficiency of *in vitro* shoot proliferation and microtuber production in potato (*Solanum tuberosum* L.), *M.Sc. Thesis*, University of Agricultural Sciences, Bangalore.

Knauss J.F. (1976). A tissue culture method for producing *Dieffenbachia picta* cv. 'Perfection' free of fungi and bacteria. *Proc. Fla. State Hort. Soc.,* **89:** 293-296.

Leifert C. and Waites W.M. (1992). Bacterial growth in plant tissue culture media. *J. Applied Bacteriology*, **72:** 460-466.

Linsmaier E.M. and Skoog F. (1965). Organic growth factor requirements of tobacco tissue cultures. *Physiol. Plant.*, **18**: 100-127.

Lloyd G. and McCown B. (1981). Use of microculture for production and improvement of *Rhododendron* spp. *Hort Science,* **15**: 416 (Abst. 321).

Maene L.M. and Debergh P.C. (1985). Problems related to *In vitro* propagated shoots. In: *In vitro techniques*: propagation and long term storage (ed. A. Schaefer Munuhr pp. 59-72. Martinus Nijhoff/Dr. W. Junk Publishers. Dordrecht.

McClelland M.T. (1989). Effects of proliferation and rooting microenvironments on micropropagated woody plants. *MS Thesis*, University of Illinois, Urbana.

Merkle S.A., Parrott W.A. and Williams E.G. (1990). Applications of somatic embryogenesis and embryo cloning. In: *Plant Tissue Culture: Applications and Limitations.* (Ed. Bhojwani, S.S.), Elsevier, Amsterdam, pp. 67-102.

Morel G and Martin C. (1952). Guerison de dahlias atteintod' une maladie a virus. *Compt. Rend. Acad. Sci., Paris* **235:** 1324-1325.

Murashige T. and Skoog F. (1962). A revised medium for rapid growth and bioassays with tobacco tissue cultures. *Physiol. Plant.*, **15**: 473-497.

Nirmala and Singh (1996). Unpublished data from thesis titled "Studies on callus induction and shoot regeneration in Anthurium". Submitted to University of Agricultural Sciences, Bangalore.

Nitsch J.P. and Nitsch C. (1969). Haploid plants from pollen grains. *Science*, **163**: 85-87.

Prasad M.C. (2000). Micropropagation studies in curry leaf (*Murraya koeniji*). Thesis submitted to University of Agricultural Sciences, Bangalore.

Prashanth K.G., Sathyanarayana B.N., Suresh N. Sondur and Deepu Methew (2002). *In vitro* shoot proliferation and callus induction in carambola (*Averrhoa carambola*). *Tropical Science.*

Raghavan V. (1986). *Embryogenesis in Angiosperms: A Developmental and Experimental Study*. Cambridge University Press, New York.

Ramachandra Madival K., Farooqui A.A., Sathyanarayana B.N., Vasundara M. and Dalia Mathew (2003). Feasibility of mass production of *Stevia rubidiana* through Tissue Culture for commercial cultivaiton. *Proceedings of global summit on medicinal plants held on September 25-30 at Mauritius.*

Ramchandra Madival K., Farooqui A.A., Sathyanarayana B.N. and Vasundhara M. (2003). Biotechnological approaches for rapid multiplication and conservation. Proceedings of global summit on medicinal plants held on September 25-30 at Mauritius.

Sathyanarayana B.N. and Jennet Blake (1993). Effect of nitrogen sources and initial pH of the media with or without beffer on *in vitro* rooting of jackfruit

(*Artocarpus heterophyllus* Lam.) Abstract of the paper published in the Proceedings First International Symposium on Adventitious Rooting - April 18-22, 1993 Texas, U.S.A. (presented in absentia).

Shackel K.A., Novello V. and Sutter E.G. (1990). Stomal function and cuticular conductance in whole tissue cultured apple plants. *J. Amer. Soc. Hort. Sci.*, **115**: 468-472.

Skirvin R.M. (1978). Natural and induced variation in tissue culture. *Euphytica*, **27**: 241-266.

Subramanya A. and Sathyanarayan B.N. (2002). Protocol for Protoplast isolation in Carrot. Unpublished data from project report entitled "callus induction of some important ornamental plant and protoplast isolation of *Daucus carota* (Carrot). Submitted by first author to Plant Tissue Culture Laboratory, Division of Horticulture, University of Agricultural Science, and Bangalore.

Vereena J. Lyngdoh and Sathyanarayana B.N. (2002). Unpublished data from thesis titled "Direct shoot regeneration from shoot tips of Gerbera". Submitted to University of Agricultural Sciences, Bangalore.

Withers L.A. (1991). Biotechnology and plant genetic resource conservation. In: *Plant Genetic Resources: Conservation and Management.* (Eds. Paroda, R.S. and Arora, R.K.), pp 273-297.

INDEX

Abscisic acid, 37
Abscission layers, 37
Acclimatization, 197
Acridine Orange, 208
Activated Charcoal, 40, 204
Adenine, 36
Adsorption, 40, 239
Adventitious shoot proliferation, 82
Adventitious shoots, 80
Adventitious somatic embryogenesis, 80
Aeroponic culture, 260
Agar, 38
Agaropectin, 38
Agarose, 38
Agrobacterium mediated transformation, 217
Agrobacterium tumefaciens, 71
Airlift Bioreactor (ALB), 259
Alcian Blue, 273
Alternate gelling agents, 39
Amino acids, 35
Androgenic haploid, 137
Anther, 74
Anther culture, 137
Antiauxin effect, 37
Antibiotic Marker, 224
Antibiotics, 40
Antioxidants, 41
Antitranspirants, 201
Applications of micropropagation, 83
Asbestos and asbestos-paper disc, 57
Ascorbic acid, 41
Aseptic transfer area, 252
Aspirator or vacuum pump, 250
ATP, 36
Autoclaves, 22
Autoclaving, 56
Automatic media dispensers, 250
Autotrophic ability, 37
Auxiliary shoot proliferation, 68
Auxins, 36
B5 medium or Gamborg's medium, 44
Bacterial artificial chromosomes (BAC), 222
Balances, 249
BAP, 37
β-glucuronidase gene (GUS), 225
Biocatalyst, 258
Bioconversion, 195
Biological delivery system, 218
Bioreactors, 257, 258
Biotransformation, 195
Blackening or browning, 91
Boron, 34
Bramhi, 92
Brown/black explant, 103
Browning, 41
Bulbils or Bulblets, 179
Calcium, 33
Callus cultures, 74
Carambola, 129
Carnation, 15
Cell culture, 74
Cell division, 37
Cell elongation, 36
Cellulase, 163
Cellulose acetate membrane, 58
Cellulose membrane filters, 58
Central zone (CZ), 81
Chilli, 116
Chlorine, 34
Citric acid, 41
Cobalt, 34

Coconut milk, 35
Coleus, 96
Compound micrography, 266
Copper, 34
Corms (cormlets), 179
Crop improvement, 70
Cryopreservation, 76, 186, 187
Cryoprotectants, 188
Cryoprotection, 188
Culture contamination, 229
Culture vessels, 27
Curry leaf, 93
Cuticle, 197
Cybridization, 173
Cymbidium orchids, 79
Cytokinin, 36
Cytokinin habituation, 242
Cytoplasm, 41
Cytoplast or cybrid, 173
DAPI, 232
Dehydration sequence, 189
Dehydration, 272
Declon Blue, 31
Delivery systems, 218
De nova shoot formation, 69
Digital pH meter, 25
Dimethyl sulphoxide, 188
Direct androgenesis, 139
Direct somatic embryogenesis, 83
Dissecting microscopes, 250
Distillation unit, 24
Dithiothreitol, 41
DNA Methylation, 215
Double haploid strains, 136
Dry heat sterilization, 56
E. coli, 223
Elicitation, 195, 196
Elicitors, 196
Electrofusion, 174
Electron donor, 41
ELISA, 232
Embryo culture, 37, 74, 149
Encapsulated shoot buds, 260
Encapsulation, 189
Endogeneous regulator, 37
Endosperm culture, 153
Ex vitro rooting, 205
Eyeball determination, 231
Filter sterilization, 57
Fixation, 272
Flame sterilization, 56
Floorplan, 248
Fluorescence microscopy, 207
Fluorescent labels, 175
Foolish seedling, 37
Frequent transfers, 239
Friable callus, 112
GA_3, 37
Gametoclonal variation, 137
Gelling agent, 40
Gelrite, 39, 40
Gene amplification, 215
Genetic transformation, 76
Genetic wipe-out, 183
Gerbera, 98, 120
Germplasm exchange, 184
Gibberella fujikuroi, 37
Gibberelins, 37
Globular stage, 123
Gloriosa superba, 127
Glutathione, 41
Gradocol membrane, 58
Ground Banana, 35
Growth of cell cultures, 37
Growth room, 250
Gynogenic haploid, 144
Gynogenic pathway, 137
Hairy root culture, 76, 194
HCl, 41
Heart stage, 123
Heater, 27
Hematoxylin, 273
Herbicide Markers, 224
Heterokaryons, 177
High Efficiency Particulate Air (HEPA), 247
Histochemical Techeniques, 271
Homokaryons, 177
Host recognition, 219
Hot air oven, 27
Hot plate, 250
Hybrid sorting, 136
Hydathodes, 199
Hydrodynamics, 258

Hyperhydricity, 39, 235
IAA, 36
IBA, 36
Immature endosperm, 155
Immersion Media, 264
In situ conservation, 183
In vitro fertilization, 74
In vitro germplasm, 72
In vitro haploid, 135
In vitro mutagenesis, 75
In vitro pollination, 74
In vitro rooting, 203
In vitro selection, 75
In vitro tuberization, 180, 181
Indirect androgenesis, 139
Indirect somatic embryogenesis, 83
Infiltration, 272
Inhibiting growth, 37
Inverted microscope, 26, 267
Iodine, 34
Iron, 34
Jackfruit, 95
Karyotypic changes, 215
Kinetin, 37
Labwashers, 250
Laminar flow hoods, 19
L-cysteine, 41
Leaching or dispersal, 239
Leaf primodium, 179
Lindemann orchid basal salt mixture, 45
Liquid nitrogen vapour, 187
Luciferase gene (LUC), 225
Macerated tissue, 260
Macerozyme, 163
Macrophotography, 266
Magnesium, 33
Magnetic stirrer, 27
Magnification, 261
MAIS, 234
Malt extract, 35
Manganese, 34
Mature stage, 123
MCPA, 36
Metabolic transformation, 195
Mevolonic acid, 37
Microfilters, 58
Microtuber, 180
Microwave oven, 250
Microchambers, 167
Micrografting, 76
Microphotography, 266
Micropropagation, 68, 79
Microscopy, 261
Microspore, 74
Microtomy, 273
Microwave sterilization, 58
Milcap, 31
Millipore system, 24
Miniature tubers, 180
Molybdenum, 34
Monoploids, 135
MS medium, 44
NAA, 36
NaOH (KOH), 41
Nemac phenomenon, 144
Nitrogen, 32
Nitsch's medium, 44
Node culture, 69
Oculars, 264
Ontogeny, 137
Operons, 220
Optical train, 261
Orange juice, 35
Orbital Shaker, 26
Organ culture, 74
Organogenesis and enhanced axillary budding, 74
Osmoticum, 162
Ovary or ovule culture, 74
Oxidation, 41
PALLAS, 232
Parafluro phenyl alanine (PFP), 144
PCPA, 36
Pectinase, 163
Peripheral zone (PZ), 81
pH meter, 250
Phenolase activity, 240
Phosphorus, 33
Photoautotrophic, 256
Phytagel, 39
Plant regeneration, 83
Pollen mother cells (PMC), 142
Polyethelene glycol, 190
Poly haploids, 135

Polyvinylpyrolidone (PVP), 239
Potassium, 33
Precursor feeding, 196
Pro-embryo, 123
Proliferation of axillary buds, 87
Protein hydrolysate, 35
Protoplast culture, 76
Protoplast fusion, 75
Protoplast Isolation, 75
Redox potential, 240
Refrigerator, 249
Reporter genes, 224
Reverse osmosis water, 24
Rhizobium, 157
Ri plasmid, 218
Rib zone (RZ), 81
Robots, 257
Sago, 40
Secondary Metabolite, 73, 191
Seed culture, 73
Selectable markers, 224
Shoot apical meristem, 81
Shoot differentiation, 37
Shoot hedges, 259
Shoot meristem, 80
Shoot necrosis, 242
Shoot tip culture, 87
Shoot tip, 82
Single node culture, 87
Sintered glass filters, 57
Solid and Liquid media, 30
Solidifying agents, 38
Somaclonal variation, 75
Somatic embryogenesis, 37, 69, 74
Sonication-assisted *Agrobacterium*-mediated transformation (SAAT), 223
Source of carbon, 37
Spontaneous fusion, 174
Stages of micropropagation, 84
Starch, 39
Sterilizing agents, 58
Stereomicrography, 266
Stereomicroscopy, 265
Stirred tank bioreactor (STB), 259
Stirrer, 250
Stochiometric efficiency, 41
Stock solutions, 46
Stomato, 198
Subliminal contaminants, 229
Sulphur, 33
Surfactants 58
Synthetic auxin, 36
Synthetic seeds/artificial seeds, 126
Tapioca, 40
T-complex, 220
Thawing, 189
Thermal Trauma, 143
Ti plasmid, 218
Tissue and organ culture, 76
Tomato juice, 35
Torpedo stage, 123
Transformation, 70
Transposable elements, 215
T-region processing, 220
T-region, 219
Tryptophan, 36
Tumor formation, 219
Tumerogenesis, 219
2, iP, 37
2,4,5-T, 36
2,4-D, 36
Types of media, 44
UV light, 59
Vir genes, 220
Vitamins, 35
Vitrification, 91, 189
Water Immersion, 263
White's medium, 44
Woody plant medium, 45
Yeast artificial chromosome (YAC), 223
Yeast, 35
Zeatin, 36, 37
Zinc, 34